Biosocial Becomings

Integrating Social and Biological Anthropology

All human life unfolds within a matrix of relations, which are at once social and biological. Yet the study of humanity has long been divided between often incompatible 'social' and 'biological' approaches.

Reaching beyond the dualisms of nature and society, and of biology and culture, this volume proposes a unique and integrated view of anthropology and the life sciences. Featuring contributions from leading anthropologists, it explores human life as a process of 'becoming' rather than 'being', and demonstrates that humanity is neither given in the nature of our species nor acquired through culture, but forged in the process of life itself. Combining wide-ranging theoretical argument with in-depth discussion of material from recent or ongoing field research, the chapters demonstrate how contemporary anthropology can move forward in tandem with ground-breaking discoveries in the biological sciences.

TIM INGOLD is Professor of Social Anthropology at the University of Aberdeen. His research is wide-ranging and interdisciplinary, spanning environment, technology and social organization in the circumpolar North, evolutionary theory, human–animal relations, language and tool use, environmental perception and skilled practice. His current work is on the interfaces between anthropology, archaeology, art and architecture.

GISLI PALSSON is Professor of Anthropology at the University of Iceland, Reykjavik. He has conducted fieldwork in Iceland, the Republic of Cape Verde and the Canadian Arctic, and has written on a variety of issues, including biomedicine, genomics, genetic history, human–animal relations, fishing, arctic exploration, environmental discourse and the history of slavery.

Biosocial Becomings

Integrating Social and Biological Anthropology

Edited by
TIM INGOLD
Department of Anthropology, University of Aberdeen

and

GISLI PALSSON
Department of Anthropology, University of Iceland

CAMBRIDGE
UNIVERSITY PRESS

CAMBRIDGE UNIVERSITY PRESS
Cambridge, New York, Melbourne, Madrid, Cape Town,
Singapore, São Paulo, Delhi, Mexico City

Cambridge University Press
The Edinburgh Building, Cambridge CB2 8RU, UK

Published in the United States of America by Cambridge University Press, New York

www.cambridge.org
Information on this title: www.cambridge.org/9781107025639

First published 2013

Printed and bound in the United Kingdom by Clays, St Ives plc

A catalogue record for this publication is available from the British Library

Library of Congress Cataloguing in Publication data
Biosocial becomings : integrating social and biological anthropology / edited by Tim Ingold, Department of Anthropology, University of Aberdeen, and Gisli Palsson, Department of Anthropology, University of Iceland.
pages cm
Includes bibliographical references and index.
ISBN 978-1-107-02563-9
1. Ethnology. 2. Physical anthropology. I. Ingold, Tim, 1948–
II. Gisli Palsson, 1949–
GN316.B55 2013
306–dc23

2012047938

ISBN 978-1-107-02563-9 Hardback

Contents

Preface

The articles in this book were developed in response to an invited panel (*Human Becomings: Beyond the 'Biological' and the 'Social'*) which we organized at the biennial meeting of the European Association of Social Anthropologists (EASA) in Maynooth, Ireland, in August 2010. When reading submitted panel abstracts for the biennial meeting and planning the event a few months before it actually took place, the EASA organizers had been struck by the absence of any proposal to seriously engage with the 'biological' and its implications for the discipline of anthropology. This seemed rather strange in the light of repeated critiques in recent years of the nature/society dualism, of the increasing frustration with received theoretical paradigms, and of growing demands for some form of integration of the social and the biological in a variety of fields and disciplines both in the humanities and social sciences (social and cultural anthropology, sociology, psychology, philosophy) and in the life sciences (biological anthropology, human genetics, evolutionary and developmental biology, environmental science). The need for integration seemed all the more pressing in the context of the study of humans, traditionally divided between the two radically separated subfields of biological-physical and social-cultural anthropology.

As a result, we were invited to organize a panel to address the theoretical dualism of nature and society and to explore possible new directions for anthropology and related disciplines. Our panel summary generated extensive interest and we received far more abstracts than we could cope with, given the time constraints of the conference. The panel itself was very well attended and generated keen interest and discussions which continue in the form of a new network that has been set up under the umbrella of the EASA: http://www.easaonline.org/networks/biosoc/index.shtml.

We would like to thank the EASA and the organizers at Maynooth who offered us a platform for the lively discussions that took place. We also thank Cambridge University Press and their anonymous reviewers who warmly embraced the concept we promised. Finally, we thank our universities for financial support.

Tim Ingold and Gisli Palsson
Aberdeen and Reykjavik

TIM INGOLD

1 Prospect

DEATH OF A PARADIGM

Neo-Darwinism is dead. The paradigm that has long dictated the terms of accommodation between the sciences of life, mind, society and culture has been brought down by the weight of its own internal contradictions, by the manifest circularity of its explanations, and by the steadfast refusal of human and other organisms to conform to the straitjacket that its architects had created for them. This is not to deny that it continues to enjoy massive public, political and financial support. Its leading protagonists are among the biggest 'names' in science. In a market-driven environment, they have become celebrities and their doctrines have become brands. They have run a propaganda machine that has been adroit in playing to popular stereotypes and ruthless in the suppression of dissenting voices, variously dismissed as ill-informed, politically motivated or temperamentally hostile to science. Some adherents of the neo-Darwinian creed have feigned puzzlement as to why so many scholars in the social sciences and the humanities refuse to sign up to it. This has been attributed, variously, to disciplinary myopia, sheer prejudice, or the allure of such fads and fashions as post-modernism, relativism and social constructionism (Perry and Mace 2010). The one possibility that adherents cannot countenance, however, is that their critics – many of whom are more widely read in the histories and philosophies of science and society than they are, and have thought long and hard about the conditions and possibilities of knowing and being in the one world we all inhabit – might have good reasons to find the paradigm wanting. To admit as much would, after all, be to question the very foundations of their own belief.

Biosocial Becomings: Integrating Social and Biological Anthropology, eds T. Ingold and G. Palsson. Published by Cambridge University Press.

Rather than seeking to counter the critical arguments that have been levelled against it, their strategy throughout has been to question the intelligence, competence and integrity of those who articulate them. This strategy marks the paradigm out as a form not of science but of fundamentalism.

In a nutshell, neo-Darwinism rests on the claim that variation under natural selection is both necessary and sufficient to explain the evolution of living things. This is not, as its advocates never tire of reminding us, a claim of genetic determinism. It does not presuppose that the units that are transmitted from generation to generation, and whose mutation, recombination and differential propagation are supposed to account for evolutionary change, are genes. The only conditions are that these units should be replicable and should encode information. When it comes to humans, for whom so much of what they do and know is ostensibly learned rather than innate, and to a lesser extent perhaps for many non-human creatures as well, it would appear that much information is transmitted cross-generationally by means other than genetic replication. Those who would integrate the human sciences into the neo-Darwinian fold have co-opted the notion of culture to refer to this informational component, arguing that its transmission attests to a second track of inheritance, running in parallel to the first track of genetic inheritance (Levinson 2009, Ellen 2010; see Palsson, Chapter 12, this volume). By analogy to genes, the replicating units of the second track have been christened 'memes'. Neo-Darwinians are themselves divided on the issue of precisely how these tracks intersect, if at all. Some, writing under the banner of evolutionary psychology, would say that the innate architecture of the human mind, shaped through the natural selection of genetically prescribed attributes under environmental conditions encountered by our most remote ancestors, strongly constrains the kinds of information that can be received, processed and passed on, and therefore imposes strict limits on the forms of transmitted culture (Tooby and Cosmides 1992, Sperber 1996). Others, keen to establish a new field of 'memetics', argue that memes can take over the mind much as a parasite can take over its host, and that they will be differentially represented in a culture to the extent that they cause the infected host to behave in ways conducive to infecting everyone else (Blackmore 2000). Either way, there appear to be two processes of evolution taking place at once, biological and cultural, by way of the variation and selection of, respectively, genes and memes (Durham 1991, Richerson and Boyd 2008).

This is the view of biology and culture, and of their co-evolution, that upholders of the neo-Darwinian paradigm like to present as on the cutting edge of science. For them, it offers the promise of a unified approach that would accommodate the entire spectrum of the human sciences under one roof. A symposium staged in London in June 2010, entitled *Culture Evolves*, purported to crown it with the unqualified approval of the scientific establishment. The meeting was one of a series of events celebrating 350 years since the founding of the Royal Society, and was co-sponsored by the Society and the British Academy. It would be hard to imagine a more high-profile or prestigious platform for launching what modern science has to say about culture and its evolution. The synopsis for the meeting read as follows:

> The capacity for culture is a product of biological evolution – yet culture itself can also evolve, generating cultural phylogenies. This highly interdisciplinary joint meeting . . . will address new discoveries and controversies illuminating these phenomena, from the roots of culture in the animal kingdom to human, cultural evolutionary trees and the cognitive adaptations shaping our special cultural nature.[1]

It is perhaps no accident that among the distinguished speakers, who included psychologists, ethologists, primatologists, archaeologists and biological anthropologists, there was not a single representative from social or cultural anthropology. For the language in which this synopsis is couched – including the divisions between biology and culture and between innate capacity and acquired content, the notion of evolution as a designer and shaper of products, and the idea (implicit in the concepts of cultural phylogeny and cognitive adaptation) that the thoughts and actions of living beings are orchestrated and controlled by programs assembled from particles of transmitted information carried around in their heads – is one that belongs, in the annals of the discipline, to a bygone era. Indeed it has long since been exposed as a sham by critical anthropologists who have drawn attention to the politics of knowledge that sets modern science and enlightened scientists over and above evolved culture and its supposedly traditional carriers. If the purveyors of this language were to take a taste of their own medicine, by treating their science as an evolved cognitive adaptation and its history as a line of phylogenetic descent, what possible credence could we attach to it?

[1] See http://royalsociety.org/events/2010/culture-evolves/.

DOES CULTURE EVOLVE?

Indeed the very first sentence of the synopsis for *Culture Evolves*, though advanced as a proposition whose truth is self-evident and beyond question, is manifestly false on three counts. First, the notion that there exists an evolved 'capacity for culture', universally present in humans in advance of the diverse content with which it is subsequently filled, is a classic example of what the philosopher Whitehead (1926) called 'misplaced concreteness' – an essentialism that fallaciously assigns a material presence, in human bodies and minds, to abstractions born of our own analytic attempts to establish a baseline of commensurability that would render all humans comparable in terms of similarities and differences. Under the guise of this capacity, evolutionary science projects onto our prehistoric forbears an idealized image of our present selves, crediting them with the potentials to do everything we can do today, such that the whole of history appears as but a naturally preordained ascent towards their realization in modernity. This is hardly a new view, having already been articulated in strikingly similar terms in the eighteenth century by thinkers of the Enlightenment whose project contemporary evolutionary psychologists, ignorant of the history of their own science, appear unwittingly to be recapitulating. Secondly, the opposition between the biological and the cultural is incoherent. It effectively reduces the biological to the innate, by contraposition to cultural forms allegedly acquired by non-genetic means, thus excluding from 'biology' the entire gamut of ontogenetic or developmental processes by which humans and other animals become skilled in the conduct of particular forms of life, while treating these skills, in so far as they vary between populations, as no more than the outward expressions of an informational supplement supplied by transmitted culture. Thirdly, and following from this, the notion of cultural phylogenies rests on an obsolete model of transmission. Linked as it is to a genealogical model that separates the acquisition of knowledge-as-information from its practical enactment, it is ill suited to describe the ways in which humans and non-humans ordinarily come to know what they do, which, as many studies have confirmed, is rather through a process of growth and guided rediscovery.

What, then, is culture? Does culture evolve? On the first score, we would say that culture is the name of a question, but it is *not* the answer. The question is: why does life, especially human life, take such manifold forms? To answer that these forms are due to culture is patently circular. The neo-Darwinian paradigm, applied to cultural as to biological evolution, is locked in this circularity. Despite much vaunted claims to the contrary, those who work within the paradigm have come up with absolutely nothing by way of an answer to the question of culture. Their procedure is rather to re-describe complex and multi-faceted, 'phenotypic' outcomes in crudely one-dimensional terms by excluding all contextually specific or so-called 'proximal' aspects that could potentially contribute to an answer, such as intentions, sensibilities, the affordances of the environment, socio-historical conditions, and the dynamics of ontogenetic development. The idea is to come up with a model of observed behaviour, a 'culture-type' (strictly analogous to the 'genotype' of biology), that is entirely context-independent. It is then supposed that this model is pre-installed inside the heads of individual carriers whence it is alleged to generate the described outcomes under the particular environmental or contextual conditions they happen to encounter. Thus, in effect, is culture 'ultimately' explained by culture. And the logical operator by which descriptions are converted into explanations, or behavioural outcomes into cognitive dispositions, is none other than variation under natural selection, here applied to culturally rather than genetically transmitted particles of information, memes rather than genes. Of course there is no denying that signs, words and ideas proliferate in the milieu wherein human lives are carried on, just as the lengths of DNA comprising the genome proliferate in the multicellular matrix within which organic forms germinate and grow. The logic of natural selection, however, requires that these signs, words and ideas, like segments of the genome, come pre-encoded with information which specifies the practices or attributes that contribute to their proliferation. This is the move that closes the loop of Darwinian explanation. Yet there is no known mechanism by which meaning can jump into minds or molecules in advance of their instillation into the life process.

Neo-Darwinian theorists have three ways of covering up the elision of *explanans* and *explanandum* entailed in this logic. One is to prevaricate over the meaning of evolution itself. At one moment it refers to changes in the relative frequencies with which allegedly self-replicating entities such as genes or memes are represented in a population; at the next to changes in manifest forms of life. Thus by a sleight of hand, it is made to appear as if having explained the one, you have explained the other. Another way is to conflate, under the concept of the 'gene' or 'meme', material instantiations (whether genomic or neural) with elements of a formal character description, commonly known as 'traits' (Moss 2003). It is this conflation that supports the illusion that segments of DNA, or their neural equivalents, encode a priori for particular practices or attributes, such that genes or memes can be said to be *for* this or that. A third way is to partition the question of how things evolve from the question of how they grow or develop, as though ontogenesis were an entirely tangential spin-off from the evolutionary process itself. Thus it appears that biological evolution is actually the evolution of the genotype, and cultural evolution the evolution of the 'culture-type'. Yet in the real world there are no genotypes and no 'culture-types'. They are models built up after the fact, constructs of retrospective analysis. It follows that neither biological nor cultural evolution – as understood within the neo-Darwinian paradigm – can occur in the world that organisms or persons actually inhabit. Such evolution can only occur in the space of abstract representations.

With this conclusion in mind, we can return to the second of the two questions posed above: 'Does culture evolve?'Clearly, in the real world, there is no such entity as 'culture' which could conceivably be said to evolve, let alone to be a product of evolution. Yet in so far as forms and practices change, over longer or shorter periods of time, there is no doubt that evolution, of a kind, does go on in this world. We could even argue that in the dynamics of this evolutionary process, and in the forms that arise within it, we can find possible answers to the question of culture: 'why does human life take so many, and such varied forms?' However this means thinking quite differently not only about culture, but also about evolution.

ON HUMAN BECOMINGS

Evolution, in our view, does not lie in the mutation, recombination, replication and selection of transmissible traits. It is rather a life process. And at the heart of this process is ontogenesis. The failure to account for the ontogenetic emergence of phenotypic form is the Achilles heel of the entire neo-Darwinian paradigm. For it has proceeded as if the form were already there, prefigured in the virtual space of the genotype or its cultural equivalent. The work of ontogenesis, then, is reduced to one of mere transcription, of the prefigured form or design into the material substrate of organic matter, or what used to be called 'protoplasm'. This way of thinking about the creation of things, whether living or artefactual, has been with us ever since Aristotle, in *De Anima*, introduced his distinction between form (*morphe*) and matter (*hyle*), arguing that the thing itself is a result of the combination of the two. This so-called 'hylomorphic' model of creation is for example invoked, for the most part quite unreflectively, whenever biologists declare that the organism is the product of an interaction between 'genes' and 'environment'. The genes are introduced into the equation as carriers of received information, which is supposed to order and arrange the formless, 'plasmic' material of the environment in the actualization of the phenotypic product. Applied to culture, the logic is just the same, and just as deep-seated in the western intellectual tradition. The only difference is that the information is carried in the virtual space of memes rather than genes – that is, in a space of ideas that are imagined somehow to have entered into people's heads, with their meanings already attached, independently and in advance of any practical involvement in the world of materials. Whether with genes or memes, the fallacy of this way of thinking lies in supposing the form miraculously precedes the processes that give rise to it (Oyama 1985). And the way to overcome the fallacy is simply to reverse the order, so as to give primacy to the processes of ontogenesis – to the fluxes and flows of materials entailed in making and growing – over the forms that arise within them. Though the solution may be simple, however, the implications are profound.

We are accustomed to thinking of ourselves as 'human beings'. The term, however, hides a paradox that is apparent as soon as we stop to ask why we do not also speak of 'elephant beings' or 'mouse beings'. Are not elephants just elephants and mice just mice? By the same token, as individuals of the species *Homo sapiens*, are not humans just humans? The catch is that humans (and elephants, and mice) can appear as such only to a mind that has already set itself on a pedestal, over and above the natural world that appears to unfold like a tapestry beneath its sovereign purview. What such a mind sees, among other things, are *human beings*. And yet in the assumption of this sovereign position, unattainable to elephants and mice, is held to reside the essence of what it means to *be human*. It is on the basis of a claim to universal humanity, defined in the first place by the possession of reason and conscience, that science authorizes its conception of human beings as comprising just another – albeit rather remarkable – species of nature. The notion of culture, then, emerges as a compromise, as the condition of beings that, while they have broken the bounds of nature, nevertheless remain encapsulated, in their thought and practice, within the constraints of received tradition. Between species of organisms and the scientists who study them, between nature and reason, human cultures figure as a middle tier in the overall scheme of things, above the former and below the latter. The very concept of the human, then, is fundamentally duplicitous: the product of an 'anthropological machine' (Agamben 2004) that relentlessly drives us apart, in our capacity for self-knowledge, from the continuum of organic life within which our existence is encompassed, and leaving the majority stranded in an impasse. To break out of the impasse, we contend, calls for nothing less than a dismantling of the machine. And the first step in doing so is to think of humans, and indeed of creatures of all other kinds, in terms not of what they *are*, but of what they *do*.

Another way of putting this, which lies at the foundations of what we attempt in this book, is to think of ourselves not as *beings* but as *becomings* – that is, not as discrete and pre-formed entities but as trajectories of movement and growth. Humanity, we argue, does not come with the territory, from the mere fact of species membership or from having been born into a particular culture or society. It is rather something we have continually to work at, and for which, therefore, we bear the responsibility (Ingold 2011: 7). Life is a task, and it is one in which we have, perpetually, never-endingly and collaboratively, to be creating ourselves. Each of us is instantiated in the world along a certain way of life or 'line of becoming' (Deleuze and Guattari 2004: 323), understood not as a corpus of received tradition but as a path to be followed, along which one can keep on going, and which others will follow in their turn. Thus unlike the incongruous hybrids of biology and culture created by the anthropological machine and conventionally known as human beings, human becomings continually forge their ways, and guide the ways of consociates, in the crucible of their common life. In so doing, they weave a kind of tapestry. But like life itself, the tapestry is never complete, never finished. It is always work in progress. Within it, we may recognize patterns, rhythms and regularities, and perhaps we might use the term 'culture' to refer to these. This is to acknowledge, however, that cultural forms arise within the weave of life, in conjoint activity. And evolution? This can only be understood topologically, as the unfolding of the entire tapestry – of the all-embracing matrix of relationships wherein the manifold forms of life that we call 'cultural' emerge and are held in place. Within this matrix, the becoming of every constituent both conditions and is conditioned by the becomings of other constituents to which it relates. These mutually conditioning relations together comprise what we can call an ontogenetic or developmental system. Forms of life, then, are neither genetically nor culturally preconfigured but emerge as properties of dynamic self-organization of developmental systems. And evolution is their derivational history.

That life unfolds as a tapestry of mutually conditioning relations may be summed up in a single word, *social*. All life, in this sense, is social. Yet all life, too, is *biological*, in the sense that it entails processes of organic growth and decomposition, metabolism and respiration, brought about through fluxes and exchanges of materials across the membranous surfaces of its emergent forms. It follows that every trajectory of becoming issues forth within a field that is intrinsically social and biological, or in short, *biosocial*. That is why we speak of humans, in this volume, not as species beings but as biosocial becomings. We admit that the terminological compression of 'social' and 'biological' into 'biosocial' is far from ideal, since the word remains tainted by connotations of hybridity and mixture, as though one could forge the human by taking a given quantum of biology and adding to it a complement devolved from a superior source in society. It has long been argued, by social and biological theorists alike, that humans – perhaps uniquely among animals – have a split-level constitution, part biological, part social, and that only by putting the two parts together can we arrive at a comprehensive account of the whole. What we intend with the 'biosocial', however, is precisely the reverse. Our claim is not that the biological and the social are complementary, or that they pertain respectively to the level of discrete individuals and to that of the wider groupings into which they are incorporated, but that there is no division between them. *The domains of the social and the biological are one and the same*. But nor is this a reductionist claim. We are not reducing the social to the biological, or vice versa. The life of a becoming (which is also, of course, the becoming of a life) could be compared to a hempen rope, twisted from multiple strands, themselves twisted from multiple fibres, each in turn twisted from its cellular and molecular constituents. It could, in principle, be examined close up or from afar, microscopically or macroscopically. But at every level of resolution we find the same complexity, the same intertwining of threads, the same metabolic exchange. Like the rope, the becoming is biological all the way up, and social all the way down.

TOWARDS A GENERAL THEORY OF EVOLUTION

The scale of the rethinking we are calling for here can scarcely be overestimated. It is not a matter of tinkering around the edges, or of adding a few more varieties of selection or tracks of inheritance, to complicate the standard neo-Darwinian picture. It is to rebuild our understanding of life and its evolution, and of our human selves, on entirely different ontological foundations. Without wishing to attach too much weight to the analogy, it is akin to the replacement of classical mechanics by the general theory of relativity. For most mundane purposes, Newton's laws of motion work well enough, since any differences between the results obtained from the application of these laws and from the principles established by Einstein would be vanishingly small. Likewise we can observe, as Darwin did (1872: 403), that while the planets have carried on in their revolutions around the sun just as Newton decreed they should, so – through the process that Darwin called 'descent with modification' – the most varied and wonderful forms have continued to evolve. But if this is to disregard the curvature of time and space brought about through gravitational mass, it is also to proceed as though every organism were a discrete entity, destined to act and react in a virtual space–time continuum in accord with its received attributes. Where for Newton the universe was a giant clock, for Darwin natural selection was a maker of watches, albeit without the intention to do so (Dawkins 1986). This mechanical conception of a clockwork world suffices as a rough approximation, so long as we keep our thinking selves well out of it. But once it is recognized that we too, in body and mind, are of the same flesh as the world, that there is no way of thinking or knowing that is not, in that sense, directed from within that which we seek to know, and that this knowing, in the practice of our science, is part and parcel of the process of becoming that makes us who we are and shapes our very humanity, this approximation is immediately exposed as the artifice it is. It is not enough to have one theory (of knowledge) for humanity and another (of being) for the rest of living nature. We need an evolutionary equivalent of the general theory of relativity that would allow our human trajectories of growth and becoming – including those of growing and becoming knowledgeable – to be re-woven into the fabric of organic life.

What follows are just some of the things that would have to be at the heart of any such theory. First, we can no longer think of the organism, human or otherwise, as a discrete, bounded entity, set over against an environment. It is rather a locus of growth within a field of relations traced out in flows of materials. As such, it has no 'inside' or 'outside'. It is perhaps better imagined topologically, as a knot or tangle of interwoven lines, each of which reaches onward to where it will tangle with other knots. This means, too, that we have either to change our understanding of the environment or to drop the concept altogether. Literally, an environment is that which surrounds. But how can a thing that knows no boundaries, that continually takes the medium into itself as it spills into the medium – or more simply, that breathes in and out – be surrounded? Indeed what we are accustomed to thinking of as an environment might better be understood as a zone of interpenetration. Within this zone, organisms grow to take on the forms they do, incorporating into themselves the lifelines of other organisms as they do so. Every organism is a site of infestation, a vast ecosystem in itself, and humans – about 90% of whose cells are actually bacteria or other micro-organisms – perhaps more than most. But the forms of organisms, as we have seen, are not already prefigured within the genome, nor are they simply transcribed into the plasmic materials of life. The only 'reading' of the genome is the process of ontogenetic development itself. In so far as the forms of organisms arise within this process, it may be described as evolutionary. The implication, however, is that the conventional division between ontogenesis and phylogenesis, or between within-generational changes that arise in the course of growth and maturation and between-generational changes in the representation of heritable attributes, can no longer be sustained. Our contention, to the contrary, is that the evolutionary process is carried forward in the life-histories of organisms themselves, along their lines of becoming.

This is to deny neither the existence of the genome, nor that differential reproduction in a space of finite resources is likely to lead to population-level changes in its composition. In this limited sense, variation under natural selection is still going on *within* the process of evolution. It is neither necessary nor sufficient, however, to *explain* the process. In a sense, the problem of explanation is the precise reverse of what it has conventionally been taken to be. It is not a question of explaining why forms change, despite being pegged down to a fixed genetic template whose constituent units are copied with remarkable fidelity across generations. It is rather one of explaining how forms remain the same, from generation to generation, in the absence of any such pegs. The more we know about the genome, the more improbable it seems that it could serve as an anchor for stability. Indeed it is hard to see how the reproducibility of organic form could be attributed to anything as fluid, as liable to getting itself tied up in knots, as prone to alteration by retro-transposition, and as susceptible to the transfer of bits and pieces back and forth with the organism's multiple and heterogeneous microbial symbionts, as the genome (Charney 2012). It is moreover evident that evolution can occur without reference to genetic change at all (whether or not such change has actually occurred), through cumulative transformations wrought through the actions of the organisms themselves on the conditions of development under which they and their successors grow to maturity. We are entirely familiar with such evolution in the field of human relations. If we do not recognize it as such, it is only because we are used to calling it history! It has been conventional to attribute evolution and history, respectively, to the two sides of humanity represented by *human beings* and *being human*: as human beings, individuals of the species *Homo sapiens*, we are said to have evolved; but in being human we are supposed to have embarked upon a process of history that has set us ever further from our biological origins. This division between evolution and history, in short, is just one more product of the anthropological machine. To dismantle the machine, as we propose, is to do away with the division, and to install in its place a general theory of the evolution of biosocial becomings.

A BIOSOCIAL SYNTHESIS

We – the editors and contributing authors of this volume – are anthropologists. Most of us are what would conventionally be called 'social' or 'cultural' anthropologists, though among our ranks are also anthropologists of a 'biological' and even 'philosophical' stripe. All of us, though, would prefer to be rid of these tiresome labels. As a discipline, anthropology has for the best part of a century been riven by internal divisions that have run along the fault line between the natural sciences and the humanities. Practitioners of physical or biological anthropology find themselves on one side; practitioners of social and cultural anthropology on the other. For some it has been too much. Academic departments have split; journals once committed to representing the entire spectrum of what was once conceived as an all-embracing science of humanity have narrowed their remit more or less exclusively to the biophysical or the sociocultural. This fragmentation, we believe, is unfortunate. It has severely weakened the discipline, and diminished its voice. Our concern is to counteract it, and to put forward a case for anthropology in the round. This is not a matter, however, of gluing the pieces back together. Our aim is rather to undo the logic that led to their division in the first place. This is the logic, as we have seen, of the anthropological machine, a machine that drove the definition of humanity as *Homo duplex*, a compound of the bio-psychological individual and the socio-cultural person. The first cracks in the complementarity thesis – namely, that by joining the individual and the person you can recreate the whole – began to appear some decades ago, with a number of studies in social anthropology that set out to show how the person is best understood *relationally*: not, that is, as a predefined position within a social structure, with its attendant rights and responsibilities, which the individual has only to assume as the actor assumes his role, but rather as a condensation of lives lived along with others. The person, according to this account, is not so much a creature of society as an active and ongoing creator of his or her own and others' selves. In the new language of relationality, these person-selves are seen as 'mutually constitutive' (Ingold 2001a).

No longer, then, could the social *persona* be regarded as a complementary 'add on' to the individual self. Rather, selves came to be understood as immanently social, in their very constitution. They are, as Gisli Palsson puts it (Chapter 2), after Marx, 'ensembles of social relations'. This insight, however, is not confined to social theorists. It is, as Palsson shows, amply born out in the knowledge and practices of indigenous peoples in many parts of the world, including the Tsimshian, Inuit and Yup'ik peoples of the northwest Pacific and Arctic coasts of North America. Nothing belongs more closely to a particular individual, or is more indexical of his or her identity, than a name. Yet among these peoples, as elsewhere, every name is itself indicative of a relation of one kind or another. The ensemble of names that someone bears corresponds to the ensemble of relations in the unfolding of which they become who they are, with their particular affections, memories, skills and sensibilities. This kind of relational thinking, however, flies in the face of the 'population thinking' that has always defined the neo-Darwinian project, according to which every individual is a discrete, bounded and enumerable entity, one of a population of such entities, and relating to other such entities along lines of contact that leave its internally specified nature unaffected. Thus the advent of relational thinking replaced the complementarity thesis with an unstable compromise: not between two *parts* of the human being, respectively social and biological, but between two completely different *ontologies* of the human, respectively relational and populational. How could both be right? The sheer incommensurability of these ontologies is largely responsible for the current deadlock in negotiations between social and biological anthropologists. To break the deadlock, we argue in this volume for a radically alternative biology. If only we could regard the organism, like the person, as an ensemble of relations – if only we could extend to biology some of the insights that have come from contemporary social anthropology – then we could open up a new synthesis in the study of biosocial relations infinitely more powerful than anything that has gone before.

To be fair to biology and to biological anthropology, the position is not as polarized as the above account would suggest. Neo-Darwinism may have caught the limelight, but it by no means commands universal assent. Many biologists and even some biological anthropologists are as adamantly opposed to its programme as we are, and for very similar reasons. This is important to emphasize because the debate is often misrepresented as for and against 'science'. Our hostility, however, is to scientism. Science and scientism are quite different. The former is a rich patchwork of knowledge which comes in an astonishing variety of different forms. The latter is a doctrine, or a system of beliefs, founded on the assertion that scientific knowledge takes only one form, and that this form has an unrivalled and universal claim to truth. One instance of scientism is the dogma that natural selection alone explains the evolution of life. Anyone who disputes this dogma is dismissed, by its more fundamentalist adherents, as anti-scientific. Yet numbered among these heretics are probably more practitioners of biological science than scholars in the humanities. Thus within the discipline of anthropology itself, the debate is not between biological anthropologists committed to science and social anthropologists who reject it; it is rather between the cult of scientism and those who are prepared to adopt a more open-ended and less complacent approach to scientific inquiry. Agustín Fuentes (Chapter 3), by training and profession a biological anthropologist, offers a shining example of how current thinking in biology is opening up ways of thinking that could expand upon, if not displace, neo-Darwinian orthodoxy. He focuses on two in particular: *niche construction* and *multiple inheritance theory*. And in Chapter 4, Eugenia Ramirez-Goicoechea takes this further with her focus on *epigenesis*, the complex, self-regulating process of life-in-the-making wherein genomic materials have the effects they do. Both authors show how particular conditions, whether organismic or environmental, may be carried on or re-produced across generations without requiring that they be 'pegged' genetically, and how these conditions may in turn be transformed through the situated activities of the organisms themselves.

INTERMINGLING LIVES

Ramirez-Goicoechea's emphasis on life as a process of making, rather than as a realization or expression of the ready-made, is key to our conception of human and other organisms as becomings. It implies, however, that metaphors of inheritance and transmission have to be treated with considerable caution. These metaphors are so deeply entrenched in the biological imagination that they are hard to shake off. Clearly, in a loose sense we can speak of conditions being 'passed on': skills, for example, may be produced anew in generation after generation of craftsmen, and farmers may continue to work the fields that their ancestors once cleared from the forest. However, in describing the former as a form of 'behavioural inheritance' (Jablonka and Lamb 2005), or the latter as a form of 'ecological inheritance' (Odling-Smee, Laland and Feldman 2003), we run the risk of disconnecting the devolution of both skills and environments from the life-process, as though skill-sets were deposited ready-made into the minds and bodies of novices, whence they have only to be acted out in life, or as though fields were but items of immoveable property rather than what they materially are: matrices of earth and crops which, if they are to bear fruit, call for continuing care on the part of those who toil in them. Thus what are often presented in the literature as parallel tracks of inheritance are, in truth, parallel and overlapping lives which, as they carry on through time, or *perdure*, also respond to one another. Parallel lives, we could say, are lived not so much in interaction as in *correspondence* (Ingold 2013). A further point of capital importance follows from this. To say that the capacities of human and other organisms are developed in and through the life process is *not*, as many critics argue, to give primacy to the environment in the determination of phenotypic outcomes, instead of to the genes. On the contrary, it is to treat the genome as an active player in the process rather than as a passive vector for the transcription of information. Together with all the other components of the developmental system, the genome is caught up in the ongoing correspondences of life-in-the-making.

In Chapter 5, Aglaia Chatjouli offers a compelling demonstration of this point. Thalassaemia is known from a biomedical perspective as a monogenic disease which impedes the normal production of haemoglobin in the blood, leading to severe anaemia and attendant health problems. No differently from everyone else, however, people diagnosed with this condition are faced with the task of keeping life going, as well as accepting the inevitability of their eventual death, amidst all the complexities and contingencies of everyday existence. For them, as Chatjouli found in her study of thalassaemics in Greece, the disease is simply a normal part of what one has to live with, as given to them as is the ground we walk and the air we breathe to us. Depending on their particular circumstances, there are myriad ways of getting by, with equally diverse outcomes that defy ready classification in terms of accepted biomedical categories. Yet these categories, belonging as they do to a strongly geneticized discourse, can directly impact on patients' life chances in so far as they affect the benefits to which they may be entitled. The thalassaemic genotype, for sufferers, is no mere abstraction: it has become part of the instituted and regulatory environment with which they have to deal. Noa Vaisman, in Chapter 6, shows likewise how it is possible, and even mandatory, to move beyond the received dichotomy between nature and nurture towards a 'third ontology', of becoming rather than being, in which who we are – our identity and humanity – is continually produced through our own actions and pronouncements. The story Vaisman tells concerns the tussle, played out in an Argentine court, between a man's regard for the parents who raised him and the claims of his birth parents, and their kin, from whom he was forcibly abducted in infancy. She describes how dissenting judges in the case sought to erase the division between 'social' and 'biological' parentage by bringing them together into a single perspective – precisely that which, in this volume, we are calling 'biosocial becoming'. In this perspective, both sets of parents would be recognized as having contributed in care and substance, albeit at different times and in different ways, to the ongoing formation of the person.

The picture is complicated, however, by another factor, concerning the proof of the man's genetic identity. Would it depend on a DNA test on a blood sample taken from his body (which he refused), or could the test legitimately be done using 'shed' DNA from bodily substances deposited through contact with his own personal effects, collected through a raid on his house? As noted above, and as this case vividly demonstrates, the materials of living bodies have a way of spilling out into the medium, where they mix and mingle with one another in that zone of interpenetration we are used to calling the 'environment'. Thus bodies may become enmeshed with one another simply by handling the same objects or by breathing the same air – or, as in the case of the thalassaemics described by Chatjouli, sharing the same blood. This intermingling of lives – that is, their sociality – is all-pervasive: it hangs in the air and runs along the ground. As Barbara Götsch observes in Chapter 7, sociality is not to be regarded as one of those 'evolved capacities' with which humans are supposed (by neo-Darwinian theorists) to come pre-equipped, but rather comprises the very relational matrix within which the evolutionary process unfolds. Through her own case study of a week in the life of a team of educationalists working for a non-governmental organization in Morocco, Götsch shows how both cognitive and technical skills are not so much transmitted as grown within communities of practice. She shows how, through joint attention in collaborative activities, team participants are able to develop a common ground of shared knowledge and experience that enables them, in turn, to follow the trails of each other's minds. In so doing, minds mingle, and their boundaries – where one mind ends and another begins – become indeterminate. But not only minds mingle; bodies do as well, for in the last resort, mind and body are indistinguishable. In going in and out of each other's minds, participants would also go in and out of each other's bodies: here the ensemble of social relations becomes *un cerveau ensemble* – a collective brain – throughout which are distributed the movements of cognition and practice that comprise the team's activity.

WHIRLS OF ORGANISMS

Long ago, the psychologist David Rubin (1988) argued that there are, in principle, two alternative ways of accounting for the reproduction of form. One is to adopt a complex structure, simple process model; the other is to adopt its converse, a model of simple structure but complex process. Though Rubin was specifically concerned with the work of memory in the reproduction of knowledge, his argument applies just as well to the reproduction of organic form. The neo-Darwinian appeal to DNA as a carrier of information exemplifies the 'complex structure, simple process' approach. It is supposed that the molecule encodes a full structural specification for the range of possible developmental outcomes which is copied into the organism at the very moment of inauguration of its life cycle, through a simple process of replication. An analogous argument, as we have seen, is adduced by those who attribute the reproduction of cultural form to the transmission of memetic rather than genetic specifications. Whether with genes or memes, a complex structure already 'copied in' to the body or mind has only to be 'copied out' in life. We argue, in this volume, for the alternative, 'simple structure, complex process' approach. No material, for example, can be more fundamental to life than water. Like other organisms, we humans depend on it, and are largely made up of it. On average, water accounts for around 60% of our body weight. With its single oxygen atom and two atoms of hydrogen, the molecular structure of water could hardly be simpler. Yet the complexity of its fluid dynamics, of its responses to disturbance, and of the forms that arise in and through its flows and circulations, are such as still to defy full understanding. We could regard the organism from a complex process perspective as a kind of eddy or 'whirl' (Cavell 1976: 52), endlessly creating itself in the current of life, just as the water of a stream, without any kind of template or central direction, forms itself into ripples, droplets and vortices.

It is in just this sense that Gaetano Mangiameli, in Chapter 8, speaks of the 'habits' of water. The people of the Kasena chiefdom of Ghana, among whom Mangiameli worked, inhabit a watery world in which it is the self-creation of things – their becoming – that lends them an aura of sanctity. The Kasena make no distinction between culture and nature. The salient distinction is rather between the *being* of ready-made or constructed things and the *becoming* of the lifeworld. To reach the sacred is to go behind the actuality of what *is* to discover the potentiality of what *may be* – the potentiality of a world of becoming, where things or ensembles make their presence felt not through what they are but through what they do. It is to look through the world of created objects to reveal the more fundamental creative process from which they have, so to speak, 'precipitated out'. Crucially, in this world of becoming there are no species in the taxonomic sense. For things to be classified as belonging to a species requires that they be excised from the flows of materials – that is, from the relations – that make them what they are. It is to convert these relations into inner attributes of which they are taken to be the effects, or in short, to revert to a world of being. The species concept, employed as a biological taxon, is a product of population thinking; it can have no purchase in an ontology that is fully relational. As Vito Laterza and his colleagues show in Chapter 9, such an ontology forces us to focus not on networks of connection between final objects, but on the meshwork of lines of material flow. These are the pathways of biosocial becoming. For materials, too, are constitutively biosocial, and we need to take them seriously. Biosocial life is a meshwork of materials endowed with properties of vitality and movement. Closely following the passage of wood through a sawmill in Swaziland, along with the movements and gestures of the men who work there and the machines they operate, Laterza finds an assembly that is throbbing with life precisely because its components are *not* perfectly joined up but rather bundled in ways that are contingent, unpredictable and potentially dangerous, and that call for continual improvisation by those whose lives are carried on both in amongst the works and far beyond.

How, then, is our humanity to be understood in this world of biosocial becomings? Far from being given unconditionally, as a baseline for activity, we have argued that humanity inheres in activity itself. It is what we do. Perhaps we should regard 'to human' as a verb. There are, then, many ways of humaning: these are the ways along which we make ourselves and, collaboratively, one another. Humans, as Istvan Praet puts it in Chapter 10, are work-in-progress. And this work calls for unremitting effort. It is not a task that can be taken up or put down at will, nor can its success or fulfilment ever be guaranteed. Although Praet's focus is on one particular indigenous group – the Chachi people of the Pacific coast of Ecuador – he shows this understanding of the human to be widespread among peoples the world over, and particularly among those who hold to an ontology of animism, that is, to an understanding of life as a creative process in which forms undergo continual generation, each in relation to the others. In such an ontology, Praet argues, humanity is both restricted and open. It is restricted to those who, through their efforts, have earned it, but it is not foreclosed – for example – by genealogical descent. This is the precise opposite of the view to which most people in western societies (including neo-Darwinian theorists) are inclined, namely that human being is restricted genealogically by species membership, but nevertheless given unconditionally rather than achieved. This leaves us, however, with the question of what happens to those who fail in the tasks of humaning. In the animic ontology, for Praet, to be beyond the pale of the human is to be beyond life itself. There is only life and non-life, or human and non-human, not different kinds of life (other species) or different kinds of human (other cultures) underpinned by the common denominators of 'nature' and 'humanity' respectively.

Yet the barrier between human and non-human, or between living and non-living, is never absolute. It can always be crossed. The risk is ever-present that one might 'fall out' of life, and this is Hayder Al-Mohammad's theme in Chapter 11. He, too, is thinking of persons as grounded in what they do rather than what they are, and is also concerned to stress the sheer precariousness of 'being-in-the-world'. It is not unconditional, and cannot be taken for granted. Drawing on the phenomenology of Heidegger, and on his own ethnographic work in the city of Basra, Iraq, Al-Mohammad considers what it might mean for a life to fall out of the meshwork. To be in the world, he contends, does not imply that one is at home in it, let alone comfortably so. It is, on the contrary, unsettling and insecure. Life and habitation both ravel and unravel, and involve as much 'falling out' – manifested in disease, loneliness, despair and ultimately death – as skilful coping. To live is to die, to be is not to be, to know is to hide from the known: to be *there* in the world is never to be *fully* there. We must ever remain enigmas to ourselves. The work of humaning, in short, holds no surety of fulfilment: it is indeed for that very reason that it can carry on. If biosocial becoming is a human predicament, it is far from an easy one!

CONCLUSION

For far too long, attempts to develop a unified approach to understanding the biological and social dimensions of human life have been frustrated by the tenacity, in the biological sciences, of a paradigm that has long since been discredited by work in social and cultural anthropology. It is not only in these subdisciplines of anthropology, however, that the neo-Darwinian paradigm has been found wanting. There have been parallel critiques in fields as various as molecular biology, epigenetics, neuroscience, ecological and developmental psychology, linguistics and the philosophy of mind. Despite diverse starting points, work in all these fields is beginning to converge on a synthesis – at once processual, developmental and relational – that is set to shatter the illusion of paradigmatic consensus perpetrated by such symposia as *Culture Evolves*. This work forces us to embark on a fundamental revision of what we mean by humanity, evolution, culture and social life, and consequently on a reconfiguration of the relations between biology, psychology and anthropology. It requires us to think of humanity not as a fixed and given condition but as a relational achievement. It requires us to think of evolution not as change along lines of descent but as the developmental unfolding of the entire matrix of relations within which forms of life (human and non-human) emerge and are held in place. And it requires us to think of these forms as neither genetically nor culturally configured but as emergent outcomes of the dynamic self-organization of developmental systems. This rethinking, we contend, amounts to a paradigm shift of a consequence for the human sciences of the twenty-first century equal to or greater than that which the Darwinian paradigm had for the sciences of the twentieth. The work that underpins this shift is going on now, and has indeed been going on for some time. Much of it, however, remains controversial or institutionally marginal to the disciplines in which it is practised, and it has still to be brought together in a way that can transform scholarship and have a significant impact on public understanding.

With this book we intend to contribute to the transformation from our vantage point in the discipline of anthropology, and in so doing, to redefine both anthropology and humanity in a way that is appropriate for our times. As several of the following chapters show, much recent questioning of the division between biological and social realms has been prompted by novel medical and biotechnological interventions. These interventions do not, in themselves, render the division invalid; for indeed, it never *has* been valid. They have however rendered its artificiality, and institutional efforts to sustain and police it, more apparent than ever before. They have, in a sense, liberated both the biological sciences and the social sciences and humanities from their old ontological moorings, allowing once divided disciplines to mix in the same melting pot. No longer does collaboration across the divide require us to set the clock back to an obsolete language of innate universals and acquired traits; instead the door is open for contemporary anthropology to move forward in tandem with groundbreaking discoveries in the biological sciences that are themselves in the throes of a paradigm shift, towards a post-genomic world wherein the rules by which neo-Darwinian logic operates no longer apply (Noble 2010, Charney 2012). Not only that, but the peoples among whom we work can also be drawn into the conversation, as wise and knowledgeable interlocutors rather than as mere carriers of evolved traditions whose only role is to provide grist to the mill of the anthropological machine. Finally, and above all, we are in a position to ground our enquiries within an ethical commitment to, and responsibility for, both our own humanity and the world in which we find ourselves. For when all is said and done, our ways of knowing are inevitably part and parcel of the generous, creative and open-ended process of biosocial becoming that is human life itself.

GISLI PALSSON

2

Ensembles of biosocial relations

> Knowledge, including (and perhaps above all) biology, is one of the ways by which humanity seeks to take control of its destiny and to transform its being into a duty. For this project, man's knowledge about man is of fundamental importance. The primacy of anthropology is not a form of anthropomorphism, but a condition for anthropogenesis.
>
> Georges Canguilhem, *Knowledge of life* (2008 [1965]: 19).

In a well-known passage in his *Theses on Feuerbach*, the sixth thesis, Marx observed (1998: 573) that 'the essence of man is no abstraction inherent in each single individual. In its reality, it is the ensemble of the social relations.' At least two points are worth noting about such a statement. For one thing, it suggests a relational, constitutive notion of the human being, an 'ensemble' firmly embedded in the company of others. A fifteenth-century term derived from medieval French (*ensemblee*), 'ensemble' denotes 'all the parts of something considered together and in relation to the whole' or 'a unit or group of complementary parts that contribute to a single effect' (The Free Dictionary 2010), one example of which would be a musical band. The other point concerns the notion of the 'social' which, for Marx, served to establish a contrast with Feuerbach, for whom the essence of humans could only 'be regarded ... as "species", as an inner, "mute", general character which unites many individuals only *in a natural* way' (Marx 1998: 573; emphasis in the original). In *Grundrisse*, Marx similarly challenged the 'illusion' of the natural individual 'posited by nature' rather than 'arising historically' (1973: 83). A growing body of scholarship uses the sketch of Marx's sixth thesis to engage with the human production

Biosocial Becomings: Integrating Social and Biological Anthropology, eds T. Ingold and G. Palsson.
Published by Cambridge University Press.

of nature in the context of environmental issues. Loftus (2009: 161), for instance, suggests that '"Society" and "nature" are . . . thoroughly interwoven as an ensemble of socio-natural relations'. What might be gained by expanding Marx's notion of the ensemble to address human becomings, by speaking of *biosocial* relations, and by collapsing the distinction between that which is posited by nature and that which arises historically? How might the life sciences of the twenty-first century, including anthropology, benefit from such an extension?

Etymologically derived from *nascere* ('to be born'), the concept of nature has connoted that which is given from birth or independent of human activities – in opposition to the 'artificial' products of human labour. Some things are provided by nature while others are constructed by humans. The naturalizing of phenomena renders them as given, elevating them to a large extent above consciousness, debate, and political action. While some schools of thought, including structuralism and evolutionary psychology, present the nature/culture axis as an essential classificatory, theoretical, and existential device, operating at the deep level of cognition, myth, language, and evolution, many scholars argue that it is neither an ethnographic nor a historical universal (see, for instance, Descola and Palsson 1996, Ingold 2000, 2011, Müller-Wille and Rheinberger 2007). Indeed for years if not decades, the nature/society divide has been subjected to critical discussion in anthropology and several other fields, including biology and philosophy. Such critique has gained increasing support as a result of growing recognition of the artificiality of nature, represented by the reconfiguring of 'life itself' and large-scale human refashioning of the global environment. This is the so-called Anthropocene, characterized by both escalating human impact and human awareness of it.

Arendt's work *The Human Condition* provides some useful insights on this score. 'For some time now', she begins, citing attempts 'to create life in the test tube', 'a great many scientific endeavours have been directed toward making life . . . "artificial", toward cutting the last tie through which even man belongs among the children of nature' (1958: 2). For Arendt, the activity of labour which 'remained stationary for thousands of years, imprisoned in the eternal recurrence of the life process to which it was tied' was finally 'liberated . . . from its circular, monotonous recurrence and transformed into a swiftly progressive development whose results have in a few centuries totally changed the whole inhabited world' (1958: 46–47). Such developments, Arendt suggested, destabilized the nature/society divide: 'The social realm, where the life process has established its own public domain, has *let loose an unnatural growth, so to speak, of the natural*' (Arendt 1958: 47; my emphasis). Arendt's work foreshadowed later understandings of the implications of biotechnology and the new genetics in highlighting the human refiguring of genomes (see, for instance, Rheinberger 1995, Rabinow 1996, Goodman, Heath, and Lindee 2003, Landecker 2007), in particular recent attempts in 'synthetic biology' to design and construct new life forms from scratch. What are the implications of the 'unnatural growth of the natural' for understanding human becomings and for the study of humans and its fragmentation?

My main concern here is with the splitting of the study of humans along the biology/society axis. While this is the central division of anthropology, separating its main tectonic plates, so to speak, it makes little sense, given the conflation of the biological and the social, failing to do justice both to the 'unnatural' growth of the natural and to some of the voices we often encounter in the field. I suggest that much depends on what is meant by the concepts of the 'biological' and the 'social' and how we see their articulation, an issue addressed by Ingold more than two decades ago in terms of a kind of 'relationships thinking' (1990: 208). As Gare argues, if the divide between nature and culture is to be bridged, 'it will be necessary to develop a science which takes becoming as basic . . . and conceives "beings" as islands of stability within the flux of becoming' (Gare 1995: 107). Attempting to move beyond both dualism and simple interactive frameworks linking separate domains of human existence, I argue that it makes sense, paraphrasing early Marx, to speak of human becomings as the configuration of ensembles of biosocial relations. The 'nature' with which we are born and which we develop is thoroughly biosocial, embodied through human activities.

In the rest of this chapter I proceed in outline as follows: The next section discusses the collapse of the theoretical notions of the biological and the social. This is followed by a discussion of the limits of gene talk and the usefulness of reimagining humans as ensembles of biosocial relations. Focusing on name talk, in the next section I argue that such reimagining resonates with a good deal of ethnography. Finally, I conclude with some general observations. Kohn reminds us (2007: 5) that while such terms as 'nature-cultures', whose current use is a 'necessary strategy', may sometimes reproduce the very dualisms we seek to overcome, they point to 'very real connections of which we need to be aware', facilitating a perspective that 'might allow us to better account for the work that goes on in the space that the hyphen seeks to bridge'. Similarly, in the absence of a better non-dualistic language, the notion of ensembles of biosocial relations may help us move beyond what Fox Keller refers to as the persistent 'mirage' of a space between nature and nurture (2010), challenging current understandings of the division of biological and social anthropology and their essentialist perspectives on key issues, including those of human nature and relatedness and the interdependencies of humans and other kinds of beings.

RETHINKING THE BIOLOGICAL: LIFE IN CONTEXT

In his heavily cited essay on the growing artificiality of life itself in the wake of the new genetics, an essay that effectively launched the concept of 'biosociality', Rabinow remarks that a 'crucial step in overcoming of the nature/culture split will be the dissolution of the category of "the social"' (1996: 99). Since the publication of Rabinow's piece, 'the social' has been scrutinized and deconstructed. While the term 'biological' has also received considerable attention it has probably remained more stable. Why should this have been the case and what might be gained by a similar dissolution of the 'biological'?

Twentieth-century biology was the culmination of a long process drawing upon several conceptual developments, including the notions of the genetic code and the cell, both of which contributed to the individuation of life. In earlier European theories of generation, Müller-Wille and Rheinberger note, 'nature and nurture, or heredity and environment, were not yet seen as oppositions' (2007: 4); the metaphors of alchemy and art were the dominant ones. Darwin and Galton, they suggest, launched a new 'epistemic space' with the application of the metaphor of *heredity*, a term (derived from the Latin *hereditas*) borrowed from the legal sphere where it was applied in the context of inheritance and succession. Anthropology positioned itself at the centre of emerging debates about this new epistemic space as 'one of the "hot spots" .... Clearly, this was a field that could not be directly accessed by experiment, the only substitute, though with its own irresolvable aporias, being the observation of "savage children"' (Müller-Wille and Rheinberger 2007: 22).

Interestingly, in his *Variation of Animals and Plants Under Domestication, Volume II*, published in 1868, Darwin referred to 'invisible characters, proper to both sexes ... and to a long line of male and female ancestors ...'; 'these characters', Darwin added, 'like those written on paper with invisible ink, lie ready to be evolved ...' (in Müller-Wille and Rheinberger 2007: 24). Continuing the textual metaphor, Galton, sometimes seen as the founding father of modern hereditary thought, used the analogy of a post office:

> Ova and their contents are, to biologists looking at them through their microscopes, much what mail-bags and the heaps of letters poured out of them are to those who gaze through the glass windows of a post office. Such persons may draw various valuable conclusions as to the postal communications generally, but they cannot read a single word of what the letters contain. All that we may learn . . . must be through inference, and not by direct observation; we are therefore forced to theorize. (Galton, in Müller-Wille and Rheinberger 2007: 6)

The metaphors of the post office, heaps of letters and illegible words of course foreshadowed the modern notion of the genetic code of the autonomous organism. While modern students of genomics are no longer 'forced to theorize' to the same degree as Galton's contemporaries, thanks to technological and digital apparatuses that allow them to gaze into what they sometimes call the 'universe within', the metaphors still draw upon the notion of the 'book of life' – indeed even more so than in the past.

The notion of the cell, another key term of modern biology, also has an interesting social history of its own. Canguilhem emphasizes (2008) that the development of the concept was intimately related to the concept of the individual. Not only, he points out, was 'cell' borrowed from the contained world of the beehive in order to represent the autonomy of the living organism, but also, unconsciously perhaps, it introduced the notion of cooperation characteristic of the construction of the honeycomb: 'Just as a honeycomb cell is an element of an edifice, bees are . . . individuals entirely absorbed by the republic. . . . It is certain that affective and social values of co-operation loom, near or far, over the development of cell theory' (Canguilhem 2008: 30).

It now seems as if a new epistemic space has been fashioned, downplaying the emphasis on individual autonomy and cooperative interaction while highlighting the mutual relationship of organism and context. Canguilhem anticipated some of the developments involved: 'would it . . . be possible', he asked, 'without rendering biology suspect, to ask of it an occasion, if not permission, to rethink or rectify fundamental philosophical concepts, such as that of life?' (2008: 59). 'The notion of milieu', he observed, echoing the *Umwelt* semiotics of von Uexküll (1982) and the dwelling perspective of Ingold (2000), 'is becoming a universal and obligatory mode of apprehending the experience and existence of living beings; *one could almost say* it is now being constituted as a category of contemporary thought' (Canguilhem 2008: 98; my emphasis). For Canguilhem, the 'individuality of the living does not stop at its ectodermic borders any more than it begins at the cell. The biological relationship between the living and its milieu is a functional relationship, and thereby a mobile one' (2008: 111).

The constitution of the category of the milieu was delayed by the successes of the new genetics in the 1950s and 60s and the more recent mapping of genomes which shifted attention from organisms to genes. Ironically, even Canguilhem himself seems to have been swayed by the rhetoric of the code of life. Now, however, as the category of the milieu is finally taking off, Canguilhem's qualification 'one could almost say' is no longer needed. The focus on milieu does not mean that the living organism has disappeared from sight, devoid of agency: on the contrary, the organism is the radiating centre of pragmatic activity: 'Biology must first hold the living to be a significative being, and it must treat individuality not as an object but as an attribute within the order of values. To live is to radiate; it is to organize the milieu from and around a center of reference, which cannot itself be referred to without losing its original meaning' (Canguilhem 2008: 113–114).

Given the embeddedness of the organism, its fleeting boundaries, the fuzzy nature of 'genomic stuff' (Palsson and Prainsack 2011) sometimes regarded as informatic assembly and sometimes as a material thing, and the co-production of organisms, species, and environments, it is difficult to see how the Aristotelian category of *zoé* – of the simple fact of living, life itself, life as such, or bare life, to mention some of the popular terms in the literature – can remain intact, as a realm separated from *bios*, the ways of life in the *polis*. 'Bare life', as Thrift remarks (2004: 147), 'is now heavily politicized'. In light of this, anthropology might be expanded and redefined as the study of more than one species – as the 'anthropology of life' – 'to encourage the practice of a kind of anthropology that situates all-too-human worlds within a larger series of processes and relationships that exceed the human' (Kohn 2007: 6). Kohn suggests that expanding the relational gaze to other kinds of beings necessitates inhabiting their multiple natures or umwelts, a transformative process of ontological blurring that he calls 'becoming' (2007: 7). In this vein, drawing on Haraway's argument (2008: 244) that 'becoming is always becoming *with*', Kirksey and Helmreich argue (2010) for a broad 'multispecies ethnography'. Indeed, it seems reasonable to broaden the discussion presented here and to regard animal becomings in general as the configuration of ensembles of biosocial relations. In this perspective, the study of humans is inseparable from the study of other animals (Palsson 2013). Many ethnographic studies would support such an argument, giving voice to people who refuse to make a fundamental difference between humans and other beings in this respect. Thus, Fuentes has analysed the relationships between macaques and humans in Bali as 'a suite of ecological, biological, and social processes that act as niche construction mechanisms' (2010: 605). Elsewhere (Palsson 2009), I have suggested it may be useful to speak of 'biosocial relations of production' to capture the different regimes and hierarchies of interspecies collaboration.

BEYOND NATURE AND NURTURE

The discovery of the double helix by the middle of the last century fostered the notion of genes as the 'secret of life', accounting for practically everything from speciation to ontogenic development, health risks and personality traits. Genes, it was assumed, kept the conversation of life going. With the development of biotechnology and the mapping of genomes, gene talk dominated the scene for years. The horizon, however, has been significantly broadened step by step, as we will see, moving from the level of single genes to large-scale environmental regimes. The failure to make significant medical predictions on the basis of single genes, apart from accounting for a few 'Mendelian' diseases, meant that analyses of genotype–phenotype correlations increasingly turned to multigenic studies assuming complex interaction and articulation. The genome, it turned out, although many laboratories still busily search for signals, did not have much to say. Moreover the stability of the genome has been seriously questioned. While it has been known for decades that cells under stress may mobilize systems that *reshape* their DNA by turning genes on and off (Jablonka and Lamb 2005: 88), the genome has largely been seen as 'an ensemble of genes strung along the chromosomes' (Barnes and Dupré 2008: 76) with identical copies in every cell. It now seems, however, that considerable variations creep in and that, as a result, 'the dogma that all the cells of an individual contain the same DNA needs revision' (Sgaramella 2010: 33). Perhaps this underlines Canguilhem's point that one must not lose sight of the radiating organism. If it turns out that the genomes of many organisms, including humans, are unstable and variable assemblies, it becomes increasingly difficult to sustain claims about authentic or 'real' genomes. It is not obvious, though, what this means for the biosocial argument developed here. Three complications of greater relevance need to be added.

For one thing, many genomes host a set of 'alien' genes, horizontally or laterally borrowed from other 'unrelated' organisms. As a result of such borrowing, the genealogical view of life that emerged in Europe during the Middle Ages – a view that drew upon the metaphor of common 'roots' – is being replaced by rhizomatic notions of relations (see, for instance, Ingold 2000, Palsson 2007a) that challenge the basic assumptions of genetic determinism, and that qualify or even undermine the verticality of established discourses of genetics, relatedness and the 'tree of life'.

Even more importantly, perhaps, growing evidence suggests that the human genome is fundamentally entangled with the microbiomes of other organisms. The human body carries with it a vast number and variety of mutually beneficial microbes, constituting about 90% of its cells and including some 99% of its genes. 'If humans are thought of as a composite of microbial and human cells', as Turnbaugh and his colleagues point out (2007: 804), and if we see 'the human genetic landscape as an aggregate of the genes in the human genome and the microbiome, and human metabolic features as a blend of human and microbial traits, then the picture that emerges is one of a human "supra-organism"'. The same applies to many other organisms. As a result, as Barnes and Dupré suggest (2008: 136), 'rather than thinking of . . . genomes as the exclusive property of individual organisms, we should think of a *metagenome* encompassing all the genomic resources available to a microbial community'.

Finally, moving beyond genes and genomes, organisms are partly regulated through a host of environmental forces that leave an imprint on their genomes that is passed on from one generation to another. Growing evidence suggests that such epigenetic regulation is prevalent in the human genome. Non-DNA related aspects of our developmental trajectory turn out to be inherited, allowing us to be the heirs of our biosocial heritage. The lives of our parents and ancestors, in other words, and the traditions and conditions of their communities in all their complexity, from dietary factors and exposure to toxic substances to behavioural habits, are embodied and memorized in our genomes, turning on some genes and silencing others, leaving a lasting 'hereditary' impact in a somewhat neo-Lamarckian fashion. Sometimes this produces severe adverse effects, including several forms of cancer. Food seems particularly important in this context. According to the growing field of nutritional epigenetics, as Landecker points out (2011: 177), 'food enters the body and *never leaves it*, because food transforms the organism's being as much as the organism transforms it. It is a model for how social things (food, in particular) enter the body, are digested, and in shaping metabolism, become part of the body-in-time, not by building bones and tissues, but by leaving an imprint on a dynamic bodily process'. Some argue that epigenetic evidence is already profoundly affecting legal and ethical discourse on genetics, equity, and justice, and that 'what is now known may only be the tip of the iceberg' (Rothstein, Cai and Marchant 2009: 22). 'The silence of the genome', as Franklin puts it, 'has given way to the cacophony of the epigenetic' (2006: 169). While the notion of epigenetics is used in different ways among biologists, social scientists and humanities scholars, it need not confuse us here. 'Biology', in any case, is far more fleeting and complex than normally imagined. And heredity and generation are biosocial things.

Years ago, Waddington (1957) launched the notion of 'epigenetic landscape' in order to move beyond simplistic genetic models of inheritance. While epigenetics is a move in the right direction, emphasizing the complexities of generation both laterally and vertically, the term itself – epigenetics – assumes an epistemic space with gene talk at the centre, juxtaposing genetics and everything else ('beyond genetics'). One of the figures Waddington presents in his book, significantly entitled *The Strategy of the Genes*, presents the 'system of interactions underlying the epigenetic landscape' (see Figure 2.1). The accompanying text underscores the 'modeling' impact of genes: 'The pegs in the ground represent the genes; the strings leading from them the chemical tendencies which the genes produce. The modeling of the epigenetic landscape, which slopes down from above one's head towards the distance, is controlled by the pull of these numerous *guy-ropes which are ultimately anchored to the genes*' (1957: 36; my emphasis). A similar genes-and-the-rest flaw is exemplified by the title of Richerson and Boyd's recent book, *Not by Genes Alone* (2008), which seeks to introduce culture into the epigenetic landscape. Dual categorizations of genes and everything else are beginning to look outdated, given the complex array of theoretical and empirical innovations nowadays associated with life itself, including those of microbiomes, 'molecular vitalism' and 'developmental systems' (see, for instance, Kirschner, Gerhart and Mitchison 2000, Oyama, Griffiths and Gray 2001), innovations that tend to characterize living regimes as ensembles of biosocial relations.

One illuminating context for research on some of the biosocial complexities discussed above is that of so-called 'extreme environments'. Irrespective of whether they are located in the Arctic or outer

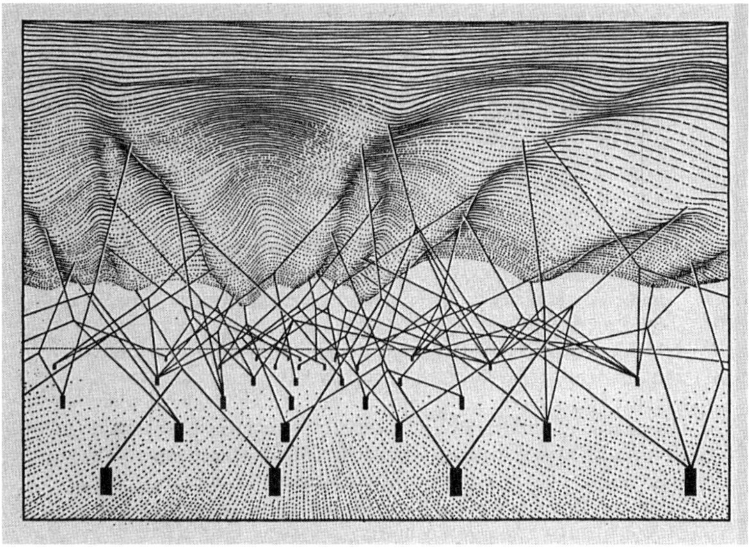

Figure 2.1. Waddington's epigenetic landscape (Waddington 1957: 36; courtesy of Taylor & Francis Books).

space, they seem necessarily to generate concerns with the constitution of the milieu and the essentials for human becoming and survival. Drawing upon her own ethnography of astronauts and Canguilhem's work on 'milieu', Olson refashions and expands the notion of biopolitics to speak of 'ecobiopolitics' as 'truth claims based on knowledge of milieu processes, power relations that take milieu as their object, and the modes of subjecthood and subjectification that designate subjects as milieu elements' (2010: 181). Space biomedicine, she argues (2010: 179), places the human species 'within a cosmic techno-ecological context of "becoming"', problematizing in the process the categories of 'life itself' and 'ecology'. While the notion of 'ecobiopolitics' grew out of a project on outer space, it is not, as Olson acknowledges (2010: 181), 'a far-out concept when put into historical context'. Along with historians and philosophers, anthropologists are exploring 'milieu's conceptual revival in today's post-genomic research in gene expression, gene regulation, and epigenetics' – in other words, rethinking the constitution of the organism and its relation to environment. It becomes increasingly difficult, in these circumstances, to maintain any kind of distinction between nature and nurture.

Just as biology has expanded its horizon from the gene to epigenomes, metagenomes, and large-scale biological regimes, linguistics has extended its discussion beyond 'language ... in and for itself', as Saussure had it (1959[1916]: 232), to the context of the speaker, cultural conventions, and discursive communities. While such extensions enhance understanding of the process of life, they have their limits. On its own, the extension of causality beyond the gene and the cell simply complicates and expands the rules of the game, much as sociolinguistics complicates and expands the rules of grammar, phonology, and syntax – the 'dictionary' in people's heads – to embrace the characteristics of events and contexts. The organism and the speaker are still rendered as if operated by codes and rules, however complex they may have become. To the extent that the metaphor of language helps to illuminate life itself, a pragmatic perspective along the lines of Malinowski and Voloshinov might offer a better way forward. As Goodwin and Duranti point out (1992: 4), a relationship of mutuality in the making of a larger whole is 'central to the notion of context (indeed the term comes from the Latin *contextus*, which means "a joining together")'. Given such a perspective, they suggest, the relationship between a speech act and context is 'much like that between "organism" and "environment" in cybernetic theory'. Context and talk, they emphasize, drawing upon Voloshinov's critique of Saussurean linguistics, 'stand in a mutually reflexive relationship to each other, ... talk, and the interpretive work it generates, shaping context as much as context shapes talk' (Goodwin and Duranti 1992: 31). Just as speakers and their utterances are inseparable from the community in which they are embedded, so the organism is inseparable from the environment.

Moss (2003) suggests such a pragmatic perspective as a way to theorize life beyond codes and genes, highlighting the roles of agency and conversations. Pointing out that much of the recent debate between gene-centrists and advocates of a new epigenesis 'can be construed as a debate about the *scope* of coding' (2003: 184), he emphasizes that the 'critical decisions made at the nodal points of organismic development and organismic life are not made by a prewritten script, program, or master plan but rather are made on the spot by an ad hoc committee [of signaling and regulatory molecules]' (2003: 186). 'After the (conflated) gene',[2] he concludes, 'it is the living organism, as an

[2] To explain the word 'conflated': Moss argues that the idea that genes 'code for' phenotypic traits is based upon an illicit conflation of two legitimate gene concepts embedded in different disciplinary practices, i.e. Genes-P, which track phenotypic markers but are indeterminate with respect to DNA sequences

active agent of its own adaptive ontogeny and evolvability, that is once again poised to move back into the ontological driver's seat' (Moss 2003: 198).

THE EPISTEMIC SPACE OF NAME TALK

The preceding discussion of biosocial relations and epigenetics is echoed in several ethnographic analyses of naming practices and name talk. Anthropologists often argue that their ethnographies need to be taken seriously, as evidence of genuine theorizing among the people with whom they study, on many of the key issues addressed by their discipline, including notions of becoming, personhood and agency. Because they are right under our nose, taken for granted, and essential to every person everywhere, personal names have often eluded the theoretical and analytical scrutiny they deserve. Focusing primarily on modern Anglo-Saxon naming practices, Finch suggests (2008: 709) that 'sociological research on names and their use is surprisingly sparse given their social significance'. Due to their central importance in our everyday lives we tend to take an ethnocentric approach to names, ignoring the variety of practices documented through ethnography and history and assuming we know all there is to know.

However, anthropology and related disciplines have created a fairly extensive literature on the variety of systems of naming in different times and cultural contexts (see, for instance, Bodenhorn and vom Bruck 2006), emphasizing that naming is a speech act shaping the life course and the person involved. The reason why names 'stick' and become powerful agents – why the speech acts work, guaranteeing what Pina-Cabral (2011) refers to as the 'ontological weight' of names – is that somehow the acting speaker is granted the licence to name by the community involved, through a formal or informal social contract. Otherwise the person would not embody his or her name, and the name would simply be discarded like worn or irrelevant clothes. Subjectivity and identity, then, are informed by the social and political environment in which naming is embedded. This is why naming practices are often a contested issue for groups campaigning for human rights and social justice.

(since these are typically based on the lack of something) and Genes-D, which are defined by nucleic acid sequences but are indeterminate with respect to phenotypic outcomes (due to the multilayered contingencies of developmental context).

One intriguing ethnographic case testifying to the significance of name talk for the understanding of human becomings is that of the Inuit of the Arctic. Throughout Inuit territory, from Alaska to Greenland, name talk has an important role to play (Bodenhorn 2000). While ethnographic interpretations do not always agree and there are significant differences between Inuit communities, the formation of an Inuk's person largely takes place through the bestowal of personal names. Names imply certain traits that are passed from one person to another, recycled with each new generation. Relatives, friends and acquaintances give each other names both as children and later in life. The set of names for a given person, as a result, is repeatedly expanded and revised during the life course. For Inuit, naming is a powerful speech act that constructs the person. The 'same' individual can be different persons depending on context and, moreover, several persons at the same time.

The role attributed by many Inuit to personal names is both similar and dissimilar to that of genes in the program theory of genes. As I have argued elsewhere:

> . . . some ethnographers have used a quasi-genetic language of 'vehicles', 'mutations', and 'substance' similar to that of mainstream genetics . . .. While, however, for many Inuit the role of personal naming is similar to that of hereditary material in modern gene talk, there are important differences. Essentially, Inuit discourse on identity and relatedness is non-reductionistic and relational, in line with the principles of epigenetics. For Inuit and other epigenetic theorists, 'biology' (in the conventional Western sense) is beside the point; fatherhood and motherhood are always 'real' and embodied. . . . It is partly through naming that children are positioned in a relational field, through which their biosociality unfolds. (Palsson 2008: 557)

Inuit name talk, then, represents an epistemic space, addressing fundamental issues of human existence, including those of human becomings, relatedness and identity.

Similarly, among the Yup'ik of Alaska the bestowal of a name signifies both belonging and identity. In the process of naming, the person becomes more than a relative: 'One gains not only social connections but a distinct social identity, becoming a unique "real person"' (Fienup-Riordan 2000: 192). The ceremony during which a child is named is called *kangiliriyaraq*, which literally means 'to provide with a beginning'. For the Yup'ik, the essence of what it means to be human passes through the name. Personhood would not be generated without parents and biological birth, but what matters above all are ancestral names defining a person's identity and position within a particular genealogy.

A further interesting and highly complex ethnographic case is that of the Tsimshian of Northwestern British Columbia analysed by Roth (2008). Roth addresses the key question of what makes a Tsimshian person, emphasizing fundamental differences between Tsimshian and white Euro-American society. For Tsimshian, reincarnation is of central importance, 'an undeniable fact of the universe .... It is a fact of nature rather than an aspect of their "culture"' (Roth 2008: 62). The English term 'reincarnation' is in fact a translation of indigenous terms denoting salmon 'running together' or 'coming back' to their spawning grounds year after year. Such togetherness and continuity is ensured through Tsimshian activities that bring names and bodies together. While much of this would apply equally to Inuit concepts and practices, Tsimshian epistemic space seems even more name-centred than that of Inuit. Thus the act of naming 'gives the person to the name' (Roth 2008: 15), as Tsimshian say, rather than the other way around. Proper names are selected from a 'basket' of 'vacant' or 'floating' names, bypassing names that have been disgraced by a wearer, 'buried' or left vacant indefinitely.

Although Inuit, Yup'ik and Tsimshian discourses are name-centred, there are important differences in emphasis. As Roth suggests (2008: 94), Northwest Coast names in general do not have the standard referencing of the Euro-American person-name format: 'These names, to the extent that they are mere names, do not refer to individuals; they *belong* to individuals and *refer* to, or rather *are*, immortal entities that ... are not souls and not quite sentient agents but are, in fact – there is no other succinct way to put it – *names*' (Roth 2008: 95). 'Tsimshian names', Roth goes on, 'refer to immortal personages – bundles of prerogatives, points or slots in a social structure ... – which are independent of the biological individual: names move from body to body during a lifetime, they can accumulate or multiply on a single body (so that a biological individual can literally *be* several personages), and successive name holders in a lineage can be linked to one another as in some sense the same social person' (2008: 97). It may be tempting to render some of these highly complicated Tsimshian terms as evidence for an essentialist nature/culture divide. For one thing, a Tsimshian name is a full-blown social actor:

> Even today, when Tsimshians orate, interact, and exchange wealth in the feast hall they do so not as the individual bodies into which they were born . . . but as names. A name is a fully formed identity – a person – with a gender, a status, a history, a future, and a living social network that links it to other names. (Roth 2008: 32)

Moreover, as Roth observes, Tsimshian name talk makes a distinction between history and structure: 'An examination of Tsimshian concepts of personhood reveals a distinction between, on the one hand, the onomastic *self*, the *essence* that is prior to any real-world social arrangements, manifested in a name rooted in a matrilineal house, and, on the other hand, the embeddedness of houses in a web of relations with other houses, a social context that is more part of *history* than of structure' (Roth 2008: 90; my emphases). While, however, the reference to 'the essence that is prior to any real-world social arrangements' might be read as a form of gene talk, naming is a biosocial process aligning persons and households through an array of human institutions and activities.

It would be wrong, then, to conclude that Tsimshian, Inuit and Yup'ik have already arrived at a kind of primitive essentialism through their naming theories. Their own form of epigenetics or developmental systems theory, in fact, moves beyond essentialism to relations and processes. Their notions of sociality and personhood evident in much of their name talk highlight the irrelevance of the idea of the autonomy of the 'biological' as commonly understood. Although Tsimshian, Inuit and Yup'ik notions of naming and kinship are anathema to genetics, they have a clear bodily reference. We may keep in mind that the *abuse* of names or harassment in the form of nicknames – a practice well documented in the ethnography (Bodenhorn and vom Bruck 2006) – also testifies to the bodily reference; often it involves physical violence in a quite literal sense, generating sensation and shame. As Scheff remarks (1988: 405), drawing upon Goffman's work on 'face', embarrassment is a firmly embodied response, involving 'a biosocial system that functions silently, continuously, and virtually invisibly, occurring *within* and *between* members of a society'.

Names not only specify and individualize their bearers, they also represent technologies of the self, serving as means of both domination and empowerment, facilitating collective action, surveillance, and subjugation – exclusion as well as belonging. While modern states and empires encourage and sometimes enforce stability of names, assuming the same name from birth to death (Scott, Tehranian and

Mathias 2002), names frequently change. Some extreme cases of renaming come from the history of slavery. Slaveholders were usually keen to rename their slaves, often with names not unlike those applied to pets and livestock. Thus, the persona of the slave was deformed with a new name, torn from its former social environment (Benson 2006: 181). Significantly, when slaves were granted freedom they often insisted upon formally receiving a new name in front of witnesses, to regain dignity and to publicly confirm the ontological weight of the new name.

Given the historical role and significance of patrilineal surnames in Europe, it need not be surprising that they still seem to provide indicators of the regional, cultural and genetic structure of continental populations. Taking a broad geographic perspective, Mateos (2007) suggests that often people's names offer a convenient window into population structures, especially in the absence of reliable knowledge about self-identified ethnicity, and, as a result, names both open up a new era of genetic genealogy and an important tool for policy in today's multicultural society. A perennial problem, however, for social and biological analysts as well as policy-makers and administrators is how to define and demarcate human 'populations'. While molecular studies removed anthropometry and the categorization of races to the sidelines decades ago, at least in scientific discourse, focusing on gene frequencies and sequences rather than phenotypic characteristics, they tend to fall back on problematic notions of populations and ethnic groups. Years ago, Ardener launched a critique of the bounded notion of populations and ethnic groups in demographic studies, a critique that seems pertinent to many modern studies of genomic differences and human variation: 'are the entities called "populations"', he asked, '*names* or *numbers*? If names: named for whom, and by whom? If numbers: counted by whom, and for whom? In asking the questions "by whom?" and "for whom?" we also ask in particular: by or for the "people" concerned? Or by or for the anthropologist or other scientific observer?' (Ardener 1989: 110; emphasis in the original). Including the human geneticist and the biological anthropologist, we might add. As many anthropologists have emphasized, among them Ardener, ethnic groups are fluid units with flexible boundaries, subject to both self-identification and naming.

The challenge is to rethink both the social and the biological, as these terms take for granted a western framework that is increasingly suspect and problematic, and, indeed, increasingly deconstructed in biology as well as in anthropology and philosophy. Kinship is both social and biological or, in other words, 'biosocial'; not however in the reductionist sense that is in common use today (as in such phrases as 'the biosociality of crime'), nor in the sense of the two separate inheritance systems of biology and culture (Richerson and Boyd 2008), but rather in the relational sense of the 'ensemble' that resonates with name talk and naming practices. Such perspectives seem to be quite broadly represented in ethnography from the Arctic to Melanesia, echoing the recent theoretical construct of the 'dividual' person. As Strathern remarks, we are 'forced to collapse the conventional analytical difference between persons and relations. Put abstractly, we could imagine persons as relations, and vice versa' (1991: 198–199; see also Bamford 2004). Relational notions of the person, however, may be closer to the old world than one might think. The historical anthropology of medieval Scandinavia developed by Gurevich, for instance, emphasizes the conflation both of persons and of persons and things and, moreover, 'a general awareness of the indivisibility of men and the world of nature' (1992: 178).

CONCLUSIONS

What are the implications, then, of the epistemic space of epigenetics, ensembles of biosocial relations, and the name talk developed by Tsimshian, Inuit, Yup'ik, medieval Scandinavians, Melanesians and many other anthropological subjects for the disciplining of anthropology and, more generally, of the life sciences? Keeping in mind the preceding discussion of name talk, meta-genomes and developmental systems, it seems pertinent that we turn anthropological boundary expertise and its observant gaze inward – to our own academic communities, to our field, its subfields and practices (Palsson 2010). Broad fields of enquiry, disciplines and subdisciplines arise, develop and (sometimes) disappear. It is important to explore this evolutionary process, how it is disciplined (in the dual sense of controlling and fragmentation), what establishes the candidacy for a field or discipline, and what languages and metaphors might be the most appropriate for the theoretical understanding of current and future developments.

Keeping in mind the biosocial relations discussed here, there are good grounds for reintegrating the two main wings of the study of humans. Given the arguments of the name talk discussed above, embodiment and materiality are not privileged themes for the natural sciences; rather they are open to useful scrutiny and theorizing right across the disciplinary spectrum. As we have seen, the Arendtian 'human condition' – the social and political life of *anthropos* – has been radically expanded and transformed; not only does the modern *polis* admit women, slaves and barbarians; a host of non-human species – animals, plants and microbes – have also entered the scene. If human becoming is best described as the configuration of ensembles of biosocial relations, a radical separation between social and biological anthropology seems theoretically indefensible. We should speak, then, of anthropology as a one-field project.

The notion of ensembles of biosocial relations, I have argued, helps to underline a few related points: humans may usefully be regarded as fluid beings, with flexible, porous boundaries; they are necessarily embedded in relations, neither purely biological nor purely social, which may be called 'biosocial'; and their essence is best rendered as something constantly in the making and not as a fixed, context-independent species-being. While naming theory has some parallels with gene talk, assuming that personhood is generated through the embodied coding of names, the parallel masks a more fundamental aspect, namely the central importance of the practices of personhood and relatedness subsumed under the activity of naming.

The empirical evidence generated by epigenetic research seems to call for a theoretical approach that abandons the rigid analytical dualism of nature and society. It is important to note, however, that nature and society have *always* been one; thus their merging is not the result of current escalations in the refashioning of life itself, nor do we need to elicit evidence of such escalations in order to demonstrate that they are inseparable. Zerilli points out that Arendt's reference to 'unnatural growth' bears an 'uncanny resemblance to what Michail Bakhtin calls the "grotesque body"... that "outgrows itself, transgressing its own limits". Indeed, this unnatural growth, this grotesque body, stands both as a reminder that nature is always already culture – what else can an unnatural nature mean? – and as an indictment to resurrect ancient borders against the body that knows none' (Zerilli 1995: 176–177). A somewhat similar notion seems conveyed in Plessner's early idea (1975) of the 'natural artificiality' of human existence. It is precisely, however, *because* of the escalations of 'unnatural growth' that we have become aware of the inseparability of nature and culture and sensitized to its implications (Szerszynski 2003). As a result, the dualism no longer sounds convincing.

Some anthropologists have attempted to bridge the nature/culture divide along neo-Darwinian lines. Thus, Richerson and Boyd stress the importance of culture in shaping human affairs, suggesting that 'culture itself is subject to natural selection' (2008: 13). For them, it is essential 'to think of genes and culture as obligate mutualists, like two species that synergistically combine their specialized capacities to do things that neither one can do alone' (2008: 194). While such an approach, in their view, would 'allow a smooth integration of the human sciences with the rest of biology' (2008: 246), it fails to clear the muddy waters. Many anthropologists would reject such an approach, not so much because they 'fear a reunion with biology', as Richerson and Boyd insinuate (2008: 14), but rather because they are uncomfortable with the evangelical commitment to 'smooth integration of the human sciences' (see, for instance, Schultz 2009). Indeed it seems that for Richerson and Boyd, the biological project, and by extension the project of the humanities, has been defined and settled for all time. 'Science', they maintain without a hint of irony, 'is *bound by its charter* to pursue explanations of human evolution!' (2008: 254; original emphasis). Such pronouncements fail to accept the fuzziness of 'biology' itself. Human becoming is a thoroughly relational, biosocial phenomenon, collective history embodied and endlessly refashioned in the *habitus*. Resisting the biologizing of kinship that pervades western discourse, epigenetics and name talk nevertheless suggest that relatedness is both biological and embodied. In such an expanded sense, biology is destiny.

We may not choose our genes in the way Inuit and Tsimshian choose kin, despite genetic engineering and modern reproductive technology. That does not mean, however, that genes are us. To reduce our 'biology' to genetic makeup, along the lines of mainstream gene talk, is to ignore the embodiment of our everyday experience (Ingold 2001a), including that of prenatal development, the intonation of language and musical sensibilities. It is difficult to see why the term 'biology' should be restricted to a fraction of what we are 'born' with. Not only would it overlook 'the relative arbitrariness of birth as a point of demarcation' (Fox Keller 2010: 75), missing the entire parenting process from conception to birth, not to mention 'labouring' itself; it would also disregard the ways in which postnatal development and becoming are outcomes of biosocial relations. An expanded notion of biology would include everything that is embodied during our development, the broad ensemble of biosocial signatures generated and assembled in the course of our lives. Such an expanded notion is equivalent to that of 'society'. Thus, the two terms have been radically merged, beyond mere 'overlap' and 'interaction'. After all, 'biology' and 'society' are not separate categories of being. As Canguilhem suggested, biological knowledge is, above all, 'one of the ways by which humanity seeks to take control of its destiny . . .. The primacy of anthropology is not a form of anthropomorphism, but a condition for anthropogenesis' (Canguilhem 2008: 19). One form of anthropogenesis is the growing industry of personal genomics, both co-produced and studied by anthropologists (see, for example, Palsson 2012).

Epistemic space, of course, does not arise from thin air. Müller-Wille and Rheinberger suggest (2007) that the modern notion of heredity was partly the product of bourgeois culture and its preoccupation with property. Moreover as we have seen, and as Canguilhem has argued (2008), the concept of the cell is inseparable from the political history of the concept of the autonomous individual. What developments in the larger world might have generated the epistemic space for developmental systems and associated theoretical constructs – and, for that matter, for name talk? While a solid answer to such questions necessitates a thorough ethnographic and historical investigation of its own, it seems safe to assume that theorizing along these lines is related to globalization and the current environmental crisis.

ACKNOWLEDGEMENTS

Research for this chapter has been supported by the University of Iceland Research Fund. I thank Carole Browner (University of California at Los Angeles), Sigurður Örn Guðbjörnsson (University of Iceland), Emma Kowal (University of Melbourne), Hannah Landecker (University of California at Los Angeles, Center for Society and Genetics), Lenny Moss (University of Exeter), Barbara Prainsack (King's College, London) and Christopher Roth (University of Wisconsin–Milwaukee) for stimulating discussions on some of the issues covered in the chapter.

AGUSTÍN FUENTES

3

Blurring the biological and social in human becomings

INTRODUCTION

Generally, evolutionary approaches to understanding humanity focus on the fossil record, on biological influences on human behaviour, or on human biological development. Equally, many social anthropologists focus on stories and experiences of being human and the trajectories and contexts of our lives as social symbolic creatures. Recently, there has been an increasing integration of perspectives that destabilize such boundaries and that focus, emphatically, on the myriad *processes* of becoming human, rather than on the state of *being* human (Dunbar *et al.* 2010, Ingold 2011, Jeeves 2011, Sussman and Cloninger 2011, Fuentes 2012). Bypassing the conceptualization of the biological and social as distinct domains of being, and thinking of them rather as intertwined processes of becoming, presents a powerful approach for anthropology. Specifically, I suggest that we need to adopt an integrative biosocial approach in order to evaluate the evolution of various human patterns (Ingold 1996, 2000, 2001a, 2007c, Fuentes 2009b, Schultz 2009). This means we should envision a myriad of entangled agents in the processes of becoming human. In this chapter I will summarize how the integration of innovative perspectives in evolutionary theory with ideas from both social and biological anthropology should provide us with more effective toolkits with which to examine patterns and processes of human becoming.

But first, we need to clarify, and move beyond, the often used and rightfully maligned dichotomy of nature and nurture in its modern form. On the one hand we have the 'social constructivist' stating that

Biosocial Becomings: Integrating Social and Biological Anthropology, eds T. Ingold and G. Palsson. Published by Cambridge University Press.

human nature is essentially, but not solely, a product of learning and socialization, and that it is socially constructed. On the other hand we have the 'Darwinian' view that human nature is essentially, but not solely, the product of natural selection. Advocates of both these perspectives hold that they best represent the real basis of why humans do what they do. Both are too simplistic to be of any use, but they remain present, although with thankfully decreasing prevalence, in academic discourse on human evolution (Fuentes 2009b, 2012, Schultz 2009, Jeeves 2011).

I want to note that only the most ardent 'Darwinians' (a misnomer actually[3]) use a simplistic mock-up of the social constructivist view as a straw man to justify their insistence that natural selection is the prime force in shaping human evolutionary pathways (Barkow *et al.* 1992, Pinker 2002, see also Fuentes 2009a, Laland and Brown 2011). While there are also some extreme social constructivists arguing against evolutionary views of humanity, this actual 'debate' is largely relegated to the most strident on each side (Segerstrale 2000). In actuality, most practitioners allow for more theoretical flexibility and overlapping of boundaries. Recently, a more nuanced interface between the evolutionary sciences and the humanistic social sciences has emerged, with a greater potential for interaction and productive entanglement (Ingold 2001a, 2007c, Fuentes 2009b, 2012, Schultz 2009, Dunbar *et al.* 2010, however, see also Mesoudi *et al.* 2010). It is from this perspective that I will elaborate my arguments.

In one sense I am the 'odd anthropologist out' in this volume, as my original training began with biological history and structure along with ecologies and then moved towards behaviour; unlike most social anthropologists who start with the behaviour and cultural context. By *biological* history, I mean the evolutionary (writ large) past of our species and lineage as understood through the analysis of the fossil record, genomics, comparative primatology and the reconstruction of past environments, ecologies and ecosystems. By *biological* structure, I mean the human physiological system conceived as a dynamic,

[3] I say this is a misnomer since current evolutionary theory is well past the basic contributions of Darwin and there is a growing pattern amongst evolutionary biologists to use the terms 'multiple processes of evolution' and 'evolutionary theory' rather than 'Darwinian evolution'. Those who tend to call themselves 'Darwinians' are often those researchers who regard natural selection as the only process of evolutionary change that matters when it comes to behaviour patterns. They also tend to be highly reductionist in their approaches (see Ingold 2001a, 2007c, Fuentes 2009a, 2009b, Laland and Brown 2011).

organic, entity. I do not see these as truly distinct and separable from human behaviour, society and cultural contexts, but a deep understanding of these elements shapes my approaches and interests in the human. My usual stance when presenting this information to biological anthropologists and human biologists is one of trying to convince them that social context and symbolic histories and schemata matter. However there is no need of that for readers of this book. Here I will focus on the evolutionary and physiological aspects, which might be more novel to less biologically inclined social scientists and anthropological humanists.

TRADITIONAL EVOLUTIONARY APPROACHES

Today we are passing a crossroads in understanding how to go about describing human evolution: a scenario in which we can switch from relatively static and reductionist approaches that focus on drives towards optimal adaptation, and move towards a multifaceted approach that leaves behind the static and focuses on ongoing process, and on adaptation that is sufficient and dynamic rather than optimal. The traditional evolutionary approach (the one generally termed 'Darwinian') envisions that humans are first evolved biologically and then exposed to an interactive overlay of social and cultural factors which in turn interact with our adapted selves, producing our *evolved nature*. The main proponents of this approach are to be found in the field known as evolutionary psychology. In this approach human behaviour, and subsequently culture, is primarily influenced by information-processing mechanisms situated in human minds that are 'elaborately sculpted product[s] of the evolutionary process' (Barkow *et al.* 1992). The human mind (the emergent functional correlate of the brain) is characterized by a set of design features (often characterized as modules) that have been constructed over evolutionary time, through the process of natural selection, to resolve adaptive problems (Barkow *et al.* 1992). Accordingly, 'evolution by natural selection provides a coherent and unified explanation of human social evolution and adaptation' (Barrett *et al.* 2002). Here it is the 'evolved information processing mechanisms' in the brain that are supposed to facilitate the human ability to 'absorb, generate, modify and transmit culture'. These evolved mental modules are the psychological mechanisms that take 'cultural' information as input and enable human behaviour to emerge as output (Barkow *et al.* 1992). Evolutionary psychology[4] is closest to the strident 'Darwinism' noted above and can best be described as 'the adaptationist program applied to the study of the human mind/brain' (see Laland and Brown 2011).

Another, slightly more flexible Darwinian approach has been proposed by human behavioural ecologists who study human behaviour with the aim of showing how it is adapted to ecological conditions. This approach seeks to understand the natural selection pressures on humans and to advance hypotheses about how the behavioural patterns exhibited today arose in response to current and past ecological pressures. Here the analysis of patterns and outcomes of human action is usually done in a piecemeal manner. Each behaviour or behaviour pattern (such as hunting, choosing mates and so on) is seen as a unit with its specific costs and benefits in regard to individual fitness.[5] Piece by piece analyses of behavioural factors provide a context for the creation of analytical models to generate testable hypotheses about selective pressures and the effectiveness of behavioural responses. These models often include a focus on the 'decision rules' or conditional strategies exhibited, dependent on ecological and social conditions (Fuentes 2009a, Laland and Brown 2011).

An underlying assumption of optimality-striving underlies both these approaches. This is the idea that in a system with multiple variants, selection will push the system towards optimal solutions (behaviours that maximize fitness). Those variants with lower fitness will eventually disappear from the system, leaving only the more successful variants. Over time a majority of the individuals in the population become optimally (or near optimally) suited to local challenges. Optimality models result in a focus on the co-variation of behaviour and socioecological environment and on behavioural variations as adaptive responses to ecological challenges. However, many modern practitioners of these approaches recognize that most systems do not operate at optimality and thus optimality models are used as yardsticks to identify other constraints in the system that might maintain or cause suboptimal responses (Sear and Mace 2008).

[4] However, more recent practitioners of evolutionary psychology seem more open to a wider range of data types and explanatory mechanisms than the originators were (see Archer 2004, Barrett *et al.* 2002).

[5] Here fitness is almost always defined by the number of offspring produced that survive to reproduce themselves: it is thus an assessment of success in getting one's own genetic makeup into subsequent generations. However, actual measures are more likely to be a proxy assessment of 'energy' expended and reproductive opportunities obtained.

While evolutionary psychologists and most neo-Darwinians assume there must be heritable genetic complexes underlying these evolved modules and adaptations, they do not seek actually to discover such genetic patterns. Rather, they assume that an ultimate linkage to DNA sequences must be involved, but accept a 'black box' intermediary step as a necessary tool to enable them to make the kinds of assertions they do about the fitness of traits and their eventual selection. It is worth noting that the field of behavioural genetics has been largely unsuccessful in linking specific genetic sequences to any trait outcomes, and that the same goes for attempts to locate specific DNA correlates of neurological and physiological systems. However, there are indications that some behavioural patterns and cognitive functions are at least relatively heritable in humans, suggesting that we may have not yet found the appropriate genomic process or more likely, that more than genetic processes of inheritance are involved (Laland *et al.* 2007, Bateson 2004, Charney 2012; and see below).

These traditional Darwinian approaches to human evolution, while at times insightful, generally result in a focus on *being* human. The assumption is that we *have* evolved and that we now need to explain the near optimality, or striving for optimality, of our behaviour, or alternatively the misalignment between our adapted patterns and the modern contexts in which we find ourselves. These perspectives also focus on the past as a baseline from which we can model both the adaptive trajectory towards modern humanity and the current behaviour that we observe. In the new approach I am suggesting here, the static, evolved, endpoint drops away and the focus becomes on how we are *evolving* bioculturally[6] in the past and present. This envisions an ongoing process of *becoming* human in the *longue durée* (that is, our evolutionary history, present and future) without the need to identify specific endpoints or optimal adaptations as anchors for our models of, and explanations for, behaviour. This is not to say that such anchors do not exist, but merely to imply that there may be one anchor, multiple anchors or no anchors at all in our evolutionary process for the particular behaviours of interest. This approach also includes a focus on *becoming* human over the lifespan, where development and enskilment are core processes to be integrated with adaptive and other

[6] Here I use the term 'biocultural' as that is the commonly used phrase in North American biological anthropology (Fuentes 2010). In European anthropologies, however, the synonym is 'biosocial' as in the title and content of this volume. In practice both terms are generally interchangeable.

systems (Ingold 1996, 2000, Fuentes 2012). The approach is not necessarily, or exclusively, rooted in explanations of any given suite of behaviours or human processes as optimal or adaptive, but at the same time it does not rule them out.

FOR EXAMPLE?

Conventional approaches to the evolution of human behaviour concentrate on the fossil record, modern human behaviour and ethnological analogy, comparative primate behaviour, structure and function in morphology, and (although not usually enough) ontogenetic development. Generally, practitioners rely on functional and optimal modelling, while treating natural selection as the principal architect of function and structure and assuming that what we are studying are 'evolved' systems (that is, that we are looking at the terminal point of a particular selective trajectory).

As an example, consider bipedalism (walking on two feet). This is usually regarded as a completed system that emerged via tight selective pressures and is best envisioned as a 'finished' adaptation in modern humans. There has been an enormous amount of speculation about this system, and Panglossian[7] functional models dominate the discourse. For example, to model the walking or running of the hominin *Australopithecus afarensis* (as a baseline for the evolution of human bipedalism), one would look at the structure of the bones of the lower limbs and feet, reconstruct the type of physical environment that *A.afarensis* was likely to have inhabited, and consider a suite of costs and benefits in terms of energetic expenditures and their relation to reproductive success. In short, one would draw up a general proposal to explain the origin of an optimal (or optimality seeking) adaptation, which has reached its final stage in modern humans. For decades, this was the basic approach to discussing the evolution of bipedalism.

However, due to fossil discoveries and analyses over the last three decades, bipedalism is no longer accorded priority as the anchor of human evolution, thus shifting the focus from bipedalism as a 'goal'

[7] This refers to the fictional philosopher in Voltaire's *Candide* and his assertion that 'all things are for the best'. Translated to an evolutionary perspective, this is one wherein traits, structures or behaviours are seen as being optimally adapted for exactly what they currently do (see Fuentes 2011: 209–212, for a summary of these proposals).

towards bipedalism as a component of a broader ecology and morphological system. It seems increasingly likely that the upper and lower body morphology that characterizes bipedalism emerged quite early on in Miocene apes that spent a significant amount of time in trees; the anatomy that we currently see as adapted to terrestrial movement has its roots in arboreal foraging and locomotion (Wood and Lonergan 2008, White *et al.* 2009, see Fuentes 2011 for a summary). In addition to this change in the 'origin' story of bipedality, current bio-behavioural research on walking and running finds that effects stemming from wearing shoes, modern activity profiles, moving on concrete or pavements, and so on, have significant impacts on the morphology and function of the feet and legs of contemporary human beings. All of this combines to cause a problem for the idea that modern human bipedality is the endpoint of an evolved system. There are multiple components involved in the process of *becoming* in legs and feet in our current world.

I suggest that there are a number of current perspectives that we can engage with and incorporate into our evolutionary questions about human behaviour as we move beyond a relatively mono-dimensional neo-Darwinism. These include taking evolving, or *becoming* in the *longue durée*, as our starting point rather than as an endpoint, and taking evolvability, dynamism, and contingency to characterize most of our genome, the structures of the body, and our behaviour. In this view, development over the life course and current and past niche construction are the co-architects, in addition to natural selection, of expressed behaviour, and multiple systems of inheritance are standard in our explanations, replacing the primacy of genetic inheritance. This results in a potentially messy and complicated, but nevertheless more accurate reliance on multiple architects and dynamic and entangled entities as opposed to static beings and environments.

How do I envision such an approach? Let's turn back to our consideration of bipedalism and feet; how do we model and think about bipedality in humans today? For a specific example, what conceptual tools might we need to deploy in order to think about the running of American sprinter Kathy Jager? In 1999, at the age of 56, at an international athletes meet in England, Jager ran the 100 meters in 13.55 seconds and 200 meters in 28.34 seconds, leaving her nearest competitors in the dust. After she won, she was asked to submit to a drugs test, and was subsequently informed that she had tested positive for anabolic steroids – performance-enhancing substances that are banned in track and field. Her times were expunged from the records, her awards were cancelled, and she was suspended from competition for two years. Jager, a grandmother of four from Phoenix, Arizona, was not taking anabolic steroids, but she was taking Estratest, a hormone-replacement drug, for menopause-related hot flushes. Estratest contains a small amount of synthetic testosterone, which is what turned up positive in the drugs test. Jager was unable to get the decision reversed, but after the suspension period ended, she resumed her competitive running career and began winning medals again.

In this case we can see that understanding the running (bipedality) of a 56-year-old woman is a good deal more complex than we might think. We cannot merely seek an ultimate explanation for the adaptive function and then assume a suite of cultural overlays. That is, it is not enough to factor in her bipedal ability, her age, and her behaviour, since that would lead us to a series of contradictory results. While, like all humans, she has the ability to run bipedally, she has also trained extensively and has thus restructured her muscles and skeletal morphology in a way atypical for her age and gender. She has also undergone the menopause and thus her endocrine patterns have shifted, but she is using a culturally mediated chemical treatment to increase certain hormones in her system which may influence her running performance and other aspects of her physiology. It is not sufficient to say that in an evolutionary sense she is 'atypical' as a runner, given her age and gender. Rather, assessing her locomotion requires a substantially integrative approach (drawing on anatomy, endocrinology and ethnography) in addition to our evolutionary perspectives on the anatomy and function of bipedality. We are not served by a simplistic or static, or even optimal, treatment of bipedalism, 'old age', gender roles, or hormone function.

Moving beyond a traditional neo-Darwinian approach requires familiarity with the emerging elements of modern evolutionary theory. In the next section I will review some key elements, with the aim of adding to our toolkit in approaching human becomings.

EMERGING PERSPECTIVES IN EVOLUTIONARY THEORY

While there are diverse threads in modern evolutionary theory, the two major recent contributions that expand on traditional Darwinian approaches are the theories of *niche construction* and *multiple inheritance*.

Niche construction theory, developed by Odling-Smee and his colleagues (2003), focuses on the interaction of individual and local ecology, on phenogenotypes (the simultaneous product of genes, structure and behaviour) and on ecosystem engineering (organisms modify, and are modified by, their own and other organisms' selective environments). Ecological inheritance (the passing down of structural and social contexts from one generation to the next) emerges as a core factor in niche construction, which includes modified selection pressures for subsequent populations in the same locations. Niche construction includes the building and destruction of niches by organisms *and* the synergistic interactions between organisms and environments. It is, according to Kendal and his colleagues,

> ... an endogenous causal process in evolution, reciprocal to the causal process of natural selection. It works by adding ecological inheritance, comprising the inheritance of natural selection pressures previously modified by niche construction, to genetic inheritance in evolution. Human niche construction modifies selection pressures in environments in ways that affect both human evolution, and the evolution of other species. (Kendal *et al.* 2011)

The process of niche construction results in a feedback dynamic such that the organisms engaged in it modify the pressures of natural selection acting on them, their descendants, and populations of other species sharing the same landscape. Ecosystem engineering results in ecological inheritance which structures the dynamic relationship between organisms and environments (niches) and thus impacts on the shape and intensity of natural selection. The niche construction approach unifies ecological, biological, and social niches rather than treating them as discrete spheres (Flack *et al.* 2006, Fuentes *et al.* 2010, Kendal *et al.* 2011). This approach is especially important when thinking about human evolution, since ecological inheritance and niche construction in general can occur by cultural means. Humans are consummate niche constructors, and bringing the niche construction perspective into attempts to understand human systems, though it adds an extra layer of complication, helps to correct more simplistic neo-Darwinian adaptationist accounts (Laland *et al.* 2007).

Human cultural processes play a particularly important part in niche construction. Kendal and colleagues (2011) point out that human ecological inheritance is exceptionally robust as it involves the social transmission and inheritance of cultural knowledge and material culture. Odling-Smee *et al.* (2003) show that human genetic inheritance, in combination with human cultural inheritance, provides for multivariate dynamics in both genetic and cultural evolutionary processes. Human behaviour results from processes of information acquisition at multiple levels, including population genetic processes, ontogenetic processes and cultural processes. Niche construction in humans emerges from all three of these processes, each of which can impact on patterns, contexts and structures of natural selection. Odling-Smee *et al.* (2003: 260–261) note that 'much of human niche construction is guided by socially learned knowledge and cultural inheritance, but the transmission of this knowledge is itself dependent on pre-existing information acquired through genetic evolution, complex ontogenetic processes, or prior social learning'.

The existence of variation in form and function, and the potential inheritance of such variations, is the central mechanism by which evolutionary changes can occur. Biologists Jablonka and Lamb (2005) argue that it is time for a 'new' new synthesis in evolutionary theory. They suggest that evolution should be redefined as the 'set of processes that leads to change in the nature and frequency of heritable types in populations over time' (Jablonka and Raz 2009: 167–168). They argue for the recognition of four primary modes of inheritance in evolutionary processes, rather than just one. They note that traditional Darwinian approaches focus on one system of inheritance only: the genetic system. Jablonka and Lamb suggest that three other inheritance systems can also have structural importance in evolutionary change. These other systems are the epigenetic, the behavioural and the symbolic. Epigenetic inheritance – that is, the inheritance of molecular or structural elements outside of the DNA – is found in all organisms. This gives rise to phenotypic variations that *do not* stem from variations in DNA but are transmitted to subsequent generations of cells or organisms (Jablonka and Raz 2009). Behavioural inheritance is the transmission, across generations, of behavioural patterns or particulars, and is found in most organisms. Finally, symbolic inheritance, the cross-generational acquisition of symbolic concepts and ideologies, is found only in humans and can have pronounced effects on behavioural patterns.

Jablonka and Lamb (2005) demonstrate that information, the variation that is the fuel for evolutionary change, is transferred from one generation to the next by many interacting inheritance systems. Variation is also constructed, in the sense that, whatever their origin, which variants are inherited and what final forms they assume depend on various 'filtering' and 'editing' processes that occur before and during transmission. Employing this perspective enables us to augment standard Darwinian approaches to human evolution which rely exclusively on behavioural or morphological proxies for genetic inheritance. Because many organisms transmit information via behaviour, acquisition of evolutionarily relevant behavioural patterns can occur through socially mediated learning. The transmission of information can occur without any linkage to genetic systems that natural selection can target. Symbolic inheritance comes with human language and the ability to engage in information transfer that can be temporally and spatially complex, contain a high density of information, and convey more-than-material descriptions. With the inheritance of symbols and symbolic thought a variety of behaviours, perceptions and beliefs can be acquired and reproduced, of potential benefit (or detriment) to human populations, without requiring any genetic correlate.

Models incorporating multiple systems of inheritance will of necessity be more complex than the generally linear genetic models of neo-Darwinian selection theory. For example, standard kin selection, the favouring of close relatives due to their genetic proximity, would have to be expanded to include social and symbolic kin, behaviours and symbols associated with kinship, marriage systems and so on, in any assessment of evolutionary processes related to kinship. This approach facilitates a more comprehensive evolutionary assessment of human systems and the resulting models may be better attuned to the actual interactions of such systems. For anthropology this perspective leads to an evolutionary concern with the way in which bodies and behavioural and symbolic systems construct, are constructed by, and interact with social and ecological niches and how, in turn, these systems interact with epigenetic and genetic systems (Fuentes 2009b). This perspective denies any hierarchy of relevance in inheritance systems and allows us to advance beyond approaches that force us to choose artificially between either a social or a biological focus.

Many elements of interest in bodies and behaviour are parts of systems and not necessarily themselves products of adaptationist selection (Oyama *et al.* 2001, Gould 2002). Niche construction and multi-inheritance theories give us evolutionary tools that incorporate these and other possibilities, and that throw light on the many explanatory and active processes that influence human becoming. These perspectives also mandate the inclusion of social histories and processes of biocultural development.

Today, in the second decade of the twenty-first century, we can envision the core of evolutionary theory as consisting of four main components: the first two comprising the traditional canon and the second two being its augmentation (from Fuentes 2012):

(1) Evolution is change over time. Specifically, it is change in genotype and phenotype across generations due to a variety of processes.
(2) Mutation generates new genetic variation. Gene flow and genetic drift move that variation around, and natural selection shapes the variation in response to environmental pressures.
(3) Niche construction theory suggests that humans and their environments are mutually interactive participants in the evolutionary process through ecological inheritance. Multiple inheritance theory proposes that evolutionarily relevant inheritance can take place at genetic, epigenetic, behavioural and symbolic levels.
(4) Our DNA alone does not determine who we are and how we behave, but it is a primary component in the development and maintenance of bodies and behaviours. Genes contain the basic instructions for the building blocks (proteins) of biological systems. Elements of the genotype and of the phenotype are connected, but not usually in a one-to-one relationship, and they are influenced by evolutionary processes.

Now that we have expanded our theoretical toolkit we can also include the concept of embodiment and the continuous process of building and shaping the human, by way of enskilment and development (Ingold 1996, 2007c). Enskilment is a process, that is: you are whom you interact with, what you eat, how you use your body, and how you perceive the world. This process of development plays out over the life course and involves physical and social development, participation in niche construction, and multiple levels of inheritance. Our local ecology, those individuals we interact with on a regular basis, our social histories, the ongoing social milieu we exist in, the way we use our bodies, all interact in a mutually malleable and entangled, non-linear manner with the parameters and trajectories of our evolutionary history and the physiological, neurological and genetic systems in our bodies, to produce a specific suite of behaviour, health and perspectives for any given person at any given time: humans are constantly becoming themselves. To conclude this chapter let me provide a few examples that help to illustrate aspects of this perspective.

EXAMPLES FROM HUMAN EVOLUTION: STRESS, PAIR BONDS AND COOPERATION

The stress response system

The hypothalamic-pituitary-adrenal axis (HPA) is a feedback loop in the body that includes the hypothalamus, the pituitary and the adrenal glands; it connects the central nervous system to the hormone systems in the body. It is a major player in stress physiology and a suite of other major neuroendocrine and behavioural functions of our body; basically it is at the forefront of the body's defences against acute stressors. It is argued that a rapid and strong HPA axis response coupled with a rapid recovery process is adaptive, providing the organism with the necessary energy to cope with a stress challenge followed by a successful return to baseline conditions (Sapolsky *et al.* 2000). There is variation between genders and between individuals in the patterns and effectiveness of HPA responses to different kinds of stressors (Kudielka and Kirschbaum 2005). A traditional evolutionary view holds that this system has evolved to perform a specific function and that differences in performance, between genders or individuals, probably reflect some form of adaptive variation. The approach I am championing here, however, suggests that the HPA system is a dynamic part of the human system and that we can better understand the way it works if we see it not as the fixed endpoint of a developmental process, responding to external cues, but rather as part of a body that continues to develop throughout its lifespan.

Obviously, fundamental components of the HPA system are determined by our evolutionary histories and the patterns of physical growth we undergo. However, the feedback loop which regulates HPA action is moderated by three pathways: via stimulation of the pituitary gland by messenger molecules; by pain and blood pressure signals travelling to the hypothalamus through the brainstem; and by psychological stimuli reaching the hypothalamus through the cerebral cortex via the limbic system. This system is shaped and affected simultaneously by neurological and endocrine pathways which combine with psychological pathways to activate, regulate, shape and alter the HPA. Thus in addition to parameters set by the interactions of genetic information and physiological development, life experiences – along with culturally and symbolically inflected perceptions of those experiences – also affect the system in its development and function. The ways in which we both impact on our hormonal states through our own behaviour (for example, in relation to diet and health), and interpret the social and physical world in judging the presence and severity of the stressors we encounter, are also important in establishing how the HPA system functions.

This means that evolutionary explanations involving the HPA must include the social contexts and experiences which have both shaped current physiologies and structures and filtered the interpretations and emotions of the individuals involved. All of these variables are affected by inheritances of multiple kinds, from DNA to nutritional context to religious beliefs and culturally derived 'common sense'. Here then, the role of the life course (including health, cultural contexts, symbolic representations, individual life histories) affects the malleability and reaction patterns of the HPA at any given time, such that changes in experiential and structural patterns in social perception and belief can be important in understanding the physiological response of an 'evolved system'. Viewing such systems simply as 'evolved' misses core aspects of their action, malleability and impact. Viewing them as becoming in the context of broader ethnographic and evolutionary processes can result in a more comprehensive, and more accurate, understanding of the 'how' and the 'why' of HPA system development, function and response.

Pair bonds and human evolution

Among primates, more species are said to have pair bonds and monogamous relationships than in any other order of mammals. It is argued that in humans the pair bond and monogamy constitute the basic social unit. However, extensive reviews of actual datasets demonstrate that primates (including humans) are not more monogamous than other mammals, but that a number of primate species live in small groups (some with and some without pair bonds), that pair bonds themselves are variable and that pair bonding is not the same as monogamy. Despite this wide variation across primates, however, it is true that in humans there is a high incidence of pair bonding, of various kinds, across the lifespan (Chapais 2008, Fuentes 2012).

There are two main types of pair bond: the social and the sexual (Fuentes 2012). The social pair bond is a strong biological and psychological relationship between two individuals that is measurably different in physiological and emotional terms from more general friendships or acquaintanceships. The sexual pair bond has a component of sexual attraction such that the paired individuals prefer mating with one another to other mating options. In all mammals pair bonds are developed by way of social interactions accompanied by the activity of neurotransmitters and a variety of hormones including oxytocin, vasopressin, dopamine and corticosterone (Ellison and Gray 2009). In non-human mammals for which the biology of pair bonds has been studied, social and sexual pair bonds frequently coincide. In humans, however, this is not necessarily the case. Humans have both social and sexual pair bonds, but the two are not always connected.

Humans, probably more than any other species, engage in extensive social pair bonding across genders and age categories. We can form pair bonds with our relatives and our closest friends; they can be with individuals of the same or different sex or age. Humans are also distinctive in forming sexual pair bonds both heterosexually and homosexually, with homosexual bonds being more common in humans than in most other primate species. Our sexual pair bonding, like our sexual activity, is not limited to contexts of reproduction (Fuentes 2012). Recent work in human evolution argues that the sexual pair bond precedes the nuclear family structure in human evolution (Chapais 2008). Indeed, Chapais (2008) argues that the sexual pair bond sets the stage for modern intensive parental cooperation. I myself have recently argued that given mounting evidence for the key role of cooperation in human evolution (Hart and Sussman 2008, Nowak and Highfield 2011), social pair bonding was a logical and effective way to enhance and expand the social networks and cooperative possibilities in human ancestral populations (Fuentes 2012).

In all human societies, social and sexual pair bonds are parts of the complex social networks that are both laid down as a core pattern in human evolution and emerge throughout the life course. There is enormous variation in the actual processes by which these bonds emerge, and in the patterns they form. The process of becoming bonded to others, both sexually and non-sexually, involves physiological and neuroendocrine systems, social and cultural histories and individual experience and context. Although pair bonding seems to be a significant part of the human evolutionary legacy, the ways in which bonding actually develops and is experienced for any individual human are remarkably contingent. Obviously, patterns and contexts of inheritance on multiple levels, alongside the social and ecological niches one moves through, and the structure and physiology of the body, all play a part. Pair bonding in humans poses a conundrum: simultaneously a significant part of our evolutionary history (perhaps an adaptation) and a major, but highly variable and idiosyncratic aspect of diverse cultures and individual lives. We can better investigate such a complex and contingent phenomenon by incorporating an approach that allows for social and physical processes of becoming alongside both traditional and recently developed tools of evolutionary analysis.

Cooperation

Despite the oft-repeated emphasis on the aggressive, warlike and competitive 'nature' of humanity, it has become clear that widespread and complex patterns of cooperation lie at the heart of human evolutionary success. There are diverse social, psychoneuroendocrinological and historical aspects to the evolution of complex cooperation in our lineage. There is support from the fossil and material record for the view that early in our evolutionary history, increasingly cooperative interactions allowed our genus to spread across the planet (Hart and Sussman 2008, Fuentes *et al.* 2010). There is also a large body of evidence demonstrating that modern human societal structures all depend on extensive cooperation. Nearly every study of human social behaviour indicates higher frequencies of, and greater emphasis on, cooperation than for any other behavioural pattern. Cooperation has evidently played a substantial role in our becoming human (Fry 2007, Nowak and Highfield 2011, Sussman and Cloninger 2011). This does not mean that humans are all egalitarian or selfless, but rather that human societies are all based on extensive and exceedingly complex systems of cooperation and mutuality, rendering any consistently selfish behavioural strategy unsustainable (Nowak and Highfield 2011).

The emergence of extensive cooperation depends on constant feedback between social and biotic ecologies, resulting in niche construction and the continual modification of selective landscapes in our evolutionary past and in our present. Nearly two million years ago members of our genus began, through physiological and behavioural means, to form increasingly complex and information-rich social networks. Coalitions, complex inter- and intra-group social negotiations and reciprocity became the primary avenues towards social and reproductive success (Dunbar *et al.*, 2010, Nowak and Highfield, 2011, Sussman and Cloninger 2011). Examining the data available for human evolution, modern human societies and human physiologies, we can see that niche construction, cooperation and conflict are intertwined in a complex system.

This system is best seen as emergent from social structures, ecologies and histories rather than as a reflection of specific adaptive patterns (Fuentes 2013). Cooperative and conflictual patterns in human behaviour result from the interactions of shared and disputed ecologies, social, economic and symbolic histories, and the perceptions of human individuals and polities, in addition to the physiological and morphological patterns of our bodies. The construction of social and historical niches interacts with features of human physiology that have emerged over our long evolutionary histories; thus it is via an ongoing process of evolutionary change and human becoming that integrations of cooperation, competition and perception have arisen.

Studying the processes and variability of cooperative interactions is extremely important in our quest to understand human evolution and behaviour. Moving away from linear or straightforward adaptationist explanations, and embracing the processes and approaches I have outlined in this chapter, can enable us to better model and interpret the human past and present.

PARTING THOUGHT

Traditional Darwinian evolutionary approaches to understanding human behaviour overemphasize adaption and optimality, characterizing humans as already 'evolved'. Their explanations are couched in terms of fitness costs and benefits for the competing variants on which natural selection acts. Social constructivists, at the other extreme, do not include human evolutionary histories and physiologies in their accounts of why we behave in the ways we do. Both of these approaches are untenable in the light of modern evolutionary theory and anthropological knowledge.

In understanding human evolution and behaviour we need to break down the boundaries between palaeontology, human physiology and studies of human biological development, social experience and symbolic life. We need to focus, emphatically, on the myriad *processes* of becoming human, rather than on the *state* of being human or of having become human ('evolved'). Instead of thinking of the processes of biological and social becoming as distinct, however, we need to see them as intertwined and integrated. This provides us with a powerful approach for anthropology in general and for the study of human evolution in particular. By integrating innovative perspectives in evolutionary theory along with new approaches in both biological and social anthropology, we will be better equipped to examine the patterns and processes of becoming human.

EUGENIA RAMIREZ-GOICOECHEA

4

Life-in-the-making: epigenesis, biocultural environments and human becomings

> In my opinion the greatest error which I have committed, has been not allowing sufficient weight to the direct action of the environment, i.e. food, climate, &c., independently of natural selection. . . . When I wrote the 'Origin', and for some years afterwards, I could find little good evidence of the direct action of the environment; now there is a large body of evidence.
>
> Charles Darwin to Moritz Wagner, 1876 (in Darwin 1887: vol. 3, 158)

FROM GENOMICS TO EPIGENOMICS: EPISTEMIC AND THEORETICAL SHIFTS

When Mendel's experiments and theories were rediscovered by Hugo de Vries (1848–1935), Darwinian evolutionary theory merged with genetics in what J. S. Huxley called the *New synthesis* and Romanes *neo-Darwinism*, because of its insistence on the primacy of natural selection. Instead of concentrating on organisms and their evolution, the new evolutionary theory was interested in genes as structural units of inheritance, selection and evolution. Genes were first defined by Johannsen in 1909 (Johannsen, 1909: 124, in Gerstein *et al.* 2007). They became the units of heredity (Noble 2010) and evolution, credited with carrying the instructions for building phenotypes from genotypes. Gene-centred evolutionary theories reduced biology to genetics and with that, life evaporated from biology (Ingold 1990). Neither the shift from focusing on individuals to focusing on populations (Dobzhansky, 1951 [1937]), nor the identification of genes with

Biosocial Becomings: Integrating Social and Biological Anthropology, eds T. Ingold and G. Palsson. Published by Cambridge University Press.

molecules of DNA (Watson and Crick 1953) changed the inherent genetic fetishism of neo-Darwinism. It was rather reinforced.

Genetic determinism remains the hegemonic theoretical perspective both in academia and in setting the biomedical research agenda (International Human Genome Sequencing Consortium 2001, Venter *et al.* 2001). It dictates the allocation of public and private research funds and has made deep inroads into the mass media, orienting the popularization of scientific knowledge and the mainstream of public culture. Nevertheless, neo-Darwinian biological and evolutionary genetic determinism has been found to be more and more problematic due to several inter-related flaws and misconceptions.

First, it is unable to account for biological complexity. Organisms cannot be reduced to DNA, nor can biological complexity be prefigured genetically (Rose *et al.* 1976). No single gene has been empirically proved responsible for the variation of a complex phenotype (Plomin and Ho 1991). Contrary to current convention, for example, there is no gene for blue eyes, but rather a genomic developmental process that disables the production of brown pigment (Moss 2001: 87). Nor is there any basis for the supposed predisposition to inherited disease such as autism spectrum disorders and Alzheimer's disease (Lock and Nguyen 2010: 339ff.). All phenotypic traits and outcomes are highly polygenic (Shao *et al.* 2008). As complex non-linear systems, biological systems do not follow the superposition principle: global changes cannot be explained by the sum of modifications of parts (*additivity*); nor is change uniformly spread throughout the system (*homogeneity*). There are no specific one-to-one correspondences between any gene for any trait but a many-to-many relation between genomic materials and phenotypic emergences. So-called genes do not code for this or that phenotype but for the impact of the environment on organisms' development (Gould 1977). Thus, selection operates at the level of the organism, the phenotype (Keller 1999). The effect of genes is indirect and manifold within multiple hierarchical (horizontal and vertical) biochemical systemic relations inside the cell. Causality is stochastic, and selection occurs within a limited range of possibilities. Life and evolution are ultimately indeterminate (Lewontin 1978).

A variety of meanings have been attributed to the concept of the gene throughout the twentieth and twenty-first centuries. Genes have been seen as discrete self-replicating units of heredity, as action programs for proteins to follow, as coded instructions, as parts of a digital sequence, as annotated entities in a database, as the subroutines of a gigantic operative system and many other things besides (Fox Keller

2005, Gerstein *et al.* 2007). The most prevalent notion is that the gene is a segment of a macromolecule called DNA that resides in the nucleus of the eukaryotic cell and carries the necessary coded information for protein production that in the end will produce a specific trait. In neo-Darwinism and genetics the gene has been mainly conceived as a digital code of instructions, the order and grammar of which can be deciphered from its signs. But in truth, DNA does not encrypt anything because there is no message. The only reading of the DNA is in the growth process of the organism itself. Void of context, the genome is an abstraction, a product of modern scientific invention (Ingold 2004: 214–215).

Evolutionary theory relies heavily on an adaptationist paradigm which, for human behaviour, is based on a rationalist theory of decision and action, namely the neoclassical economic paradigm of self-interest and methodological individualism. It is supposed that if they are to survive, organisms have to adapt to a pre-existent reality ('nature') that imposes itself onto all living and non-living things as an external environment, as a *world-out-there*. At its most basic, this world is conceived spatiotemporally as a generic and empty environment for different forms of life to inhabit, prior to its differentiation into the kinds of niches in which organisms could realistically develop and evolve (Griffiths 2007). Nature, then, is supposed to challenge the individual (or a population) by posing problems to be solved, with a limited range of resources. Yet there is no way of determining a priori what a resource can be for an organism. Challenges and resources are defined locally, depending on many inter-related proximal and distal factors. Therefore, there can be no common environment for all kinds of organisms, least of all for social ones, given their varied ethnographic and historical ways of eco-relating with other species, objects and themselves (Descola 1996, Palsson 1996). Even organisms, dead or alive, can be environments for other organisms, as for example whale carcasses that may feed marine worms for more than twenty years (Wiklund *et al.* 2009).

Gene-centred biology, evolutionism, ecology and cognitivism – as well as fields that take cognitivism for granted such as behavioural ecology (Standen and Foley 1989), sociobiology (Hamilton 1964, Trivers 1971, Wilson 1975), memetics (Dawkins 1982), evolutionary psychology (Buss 1994, Cosmides and Tooby 1994, Pinker 2002) and neo-Darwinian cognitivism – have systematically ignored development and ontogeny. Johannsen's (1911) reinterpretation of Weismann's germ plasma theory established as dogma the non-reversibility of the relation between genotype and phenotype, eliminating any possible evolutionary consequences of behaviour, action and ontogeny. The genotype (germinal plasma) was the only thing that descendants would inherit from their ancestors: acquired qualities could not be inherited. Development and the phenotype were (and are) reduced to the combined effects of interaction between the genetic program and external environmental factors.

In line with the pronounced dualism of the western epistemological tradition, neo-Darwinian evolutionists, biologists and cognitivists have continued to endorse a clear-cut division between nature and culture, giving pre-eminence to the first term in their conception of the so-called 'life sciences'. Human behaviour and culture – the 'extended phenotype' (Dawkins 1976, 1982) – are naturalized through the application to them of the paradigms of genetic determinism and natural selection.

In the 1980s a more systemic and dynamic perspective on biology proposed a less reductionist view of the genome. It was found that the enumeration and description of the human genome (International Human Genome Sequencing Consortium 2001, Venter *et al.* 2001) could not offer a complex enough account of how biological processes are started up and work in an interlinked fashion (ENCODE 2004, Fox Keller 2005).[8] On the one hand, the bounded concept of the gene began to be challenged. We cannot tell where a gene starts or ends. DNA segments splice, shuffle and reshuffle, swap and shift places, as a condition for protein production (Pearson 2006). They recombine and interact dynamically with other genomic, cellular, histological, hormonal, physiological and anatomical network systems. Nearly 43% of the

[8] In 2003 the ENCODE (Encyclopedia of DNA Elements) Consortium was founded, now encompassing more than 81 research institutions. ENCODE has provided a huge corpus of research and knowledge. The project 'provides information on the human genome far beyond that contained within the DNA sequence – it describes the functional genomic elements that orchestrate the development and function of a human. The project contains data about the degree of DNA methylation and chemical modifications to histones that can influence the rate of transcription of DNA into RNA molecules . . . [It] also examines long-range chromatin interactions, such as looping, that alter the relative proximities of different chromosomal regions in three dimensions and also affect transcription. Furthermore, the project describes the binding activity of transcription-factor proteins and the architecture (location and sequence) of gene-regulatory DNA elements, which include the promoter region upstream of the point at which transcription of an RNA molecule begins, and more distant (long-range) regulatory elements. . . . ENCODE catalogues the sequences and quantities of RNA transcripts, from both non-coding and protein-coding regions' (ENCODE 2004, see also 2007).

sequences that compose the human genome are transposed from one place to another. Dynamic processes of transposable and retrotransposable elements, copy number variations, and so on, suggest that genomic DNA variability is much higher than previously thought (Charney 2012: sections 4.1 and 6).

Nearly 98.5% of DNA does not code for any proteins. Composed mainly of repetitive sequences, and because of its unknown function, it was previously called *junk* DNA. However, 80% of DNA has important biochemical functions, particularly outside the protein-coding regions (ENCODE 2012). These non-protein-coding transcripts play an important role in the complex organization of interleaved protein-coding and non-protein-coding transcripts that calls into question the currently assumed physical boundedness of the gene (Gingeras 2007). This DNA plays a decisive role as a regulatory reservoir, also contributing to other fundamental aspects of molecular structuring such as repairing damaged DNA, and sometimes even becoming the main active component beyond the coding sequence itself (Shapiro and Sternberg 2005). The standard view of the RNA molecule as only a transcriptional factor has also been empirically questioned: it may not only produce proteins during development thanks to modified small mRNA transcripts, but also play a part in *interfering* gene expression (RNAi).

Hence, genomic materials cannot be specified independently of their dynamicity within a whole, hierarchically organized network of regulatory systems that interact with each other at specific times and places within the cell (Pardini and Guimaraes 1992, El-Hani 2007). The current emphasis is on what kinds of biological function these sequences intervene in and when and how they do it. We could think then of a *reactive genome* which is activated, stimulated and partly regulated by the dynamic interactions of cell activity and its environmental signals (Shapiro 2005: 93). Cells regulate the genetic engineering of genome system architecture by specifying and demarcating its products; their specific linkages and cooperativity allow for the integration of responses and reactions. As the immunological system shows, the cell's natural genetic engineering enhances the efficiency of the search for genome configurations that encode functionally complex systems, and supports the hierarchy of system architectures, even taking control in DNA restructuring (Shapiro and Sternberg 2005: 96–7). At the level of the cell, physiology cannot be separated from genomic regulation (Shapiro 1999, Stotz and Griffiths 2008).

The time has come for a new conceptual description of the gene. Since these biological processes and their outcomes are stochastic, probable but within constraints (Theise and Harris 2006), such that we cannot predict with certainty the role of genes without taking into account the complex regulatory dynamics in which they are enmeshed, combinatory, diffuse and soft logic models, and semiotic and linguistic metaphors, seem more apt. We need a gene concept that allows for self-organization, the mutual historical constitution of parts and whole, the global dynamics and emergent properties of the cell, signals and networks of interconnection and intercommunication, the relative coherence of biological processes between contingency and determinacy, and a new lexicon that favours interdisciplinary understanding (Fox Keller 2005). Gerstein *et al.* (2007: 676) suggest an operative definition of the gene that should include the following conditions: that it is retrospectively compatible with the phenotypic field with which it has been traditionally linked; that it can be applied to any organism; that it is simple and understandable, specific and practical for analysis and description; and last but not least, that it is consistent with phenomena and processes and their nomenclatures. Accordingly, they argue, 'the gene is a union of genomic sequences encoding a coherent set of potentially overlapping functional products' (Gerstein *et al.* 2007: 678, see also Moss 2001: 88–89).

The major shift in the biomedical research agenda brought about by this transition to non-linear and complex-systems thinking cannot be ignored. It is a shift from structural genomics to functional genomics, from genomics to proteomics, from regarding disorders as monogenic to recognizing their multifactorial causes, from gene action to gene regulation and the analysis of multiple genes in gene families, pathways and systems, and from mutations to pathogenesis (Peltonen and Mckusick 2001).

ONTOGENY AND DEVELOPMENT: TOWARDS EPIGENESIS

Ontogeny is the process of development that takes place throughout an organism's life course (Robertson 1996). The capacities and activities of organisms are emergent properties of developmental systems. *Developmental systems theories* (DST) (Oyama 1985, 1992, Oyama *et al.* 2001) have contributed decisively to our rethinking of life processes and their dynamics in terms of ontogenetic development. Likewise the relatively new interdisciplinary field known as *developmental sciences* has contributed to our understanding of social and political aspects of the life course (Magnusson and Cairns 1996). Developmental processes are generative and comprise the relational fields within which interactions and forms emerge (Ingold 1991, Lewontin 1983). They are the outcomes of previously embodied histories of regular interactions and experiences that have sculpted the organism's life and, in so doing, opened up further paths of stability and change (Lehrman 1953).

A shift towards considering organisms' properties as developmental outcomes – at multiple hierarchical levels – is becoming unavoidable in fields ranging from post-genomics, evolutionary developmental biology (*Evo-Devo*) and evolutionary epistemology (*EE*) to evolutionary thinking in general (Müller and Newman 2005: 487). If phylogeny is about continuity and change in respect to other species, ontogeny is about continuity and change within the organism's own species and its discontinuity with others. It is in the organism that ontogeny and phylogeny are mutually specified (Gould 1977). Ontogeny creates life and evolution, because all changes take place within developmental processes (Garstang 1922, cited in Hall 2000). Genetic activity is neither independent of, nor external to, the organism's developmental system (Gottlieb 2001:47–8). Development is not a factor of biology or of evolution but the very same condition for both; it creates certain possibilities for evolution to act upon.[9] The organism as a developmental system becomes once again the scenario of life and evolutionary change, as a multiplex dynamic field of and for action (Bateson 2005, Mayr 1999).

Epigenesis is the crucial concept from which the argument of this chapter stems, regarding the biosociocultural and political constructions of our local developmental trajectories. If epigenesis originally referred to the modulation of non-DNA-based genetic activity, now its meaning covers a wider range of more complex processes. Epigenesis is the directional molecular process of genetic activation, expression, revelation, suppression and regulation (Monod 1970, Ho and Saunders 1979). It influences what, how, where and when genomic materials are silenced or expressed, through the guidance and regulation of genomic products (that is, regulatory genes) in a vast system of developmental regulation.

[9] An interesting example of a development-producing structure that, in its turn, constrains further development, is that of dental disposition in mice. The first molar is bigger than the following one. Likewise, the second is bigger than the third. Molar development follows an inhibitory model of succession and indirectly sets in train other changes which may impact in specific ways, such as changes in diet, which may, in turn, induce other stochastic evolutionary consequences (Kavanagh *et al.* 2007).

Mainly but not only by methylation and histone acetylation, access to DNA for transcription, protein and enzymatic activity and transformation is either opened up or closed. Integrated regulatory mechanisms involving environmental signals, and their combination with specific activational factors, determine which sequence will be recruited for transcription, when and where, including its beginning and ending, what coding and non-coding parts will be spliced, how the remaining parts will be reassembled and, if so, translated, what nucleotides will be inserted, substituted or deleted, and what novel nucleotide information will be produced in the sequence of genomic products (Stotz 2006). The ensemble of transcriptional factors and their binding events that specify the on-and-off states of genes in a combinatorial fashion constitute the cells' regulatory network (ENCODE 2012: 91).

Epigenesis intervenes in the cellular division of the egg, the specialization of pluripotent cells for physiological processes, tissue construction and growth, and other biological operations of further hierarchical complexity. From a single cell (zygote), patterns of silencing and amplification of genome products produce the nearly two hundred different stem cells that lead to specialization and functionality through cell diversification. Cellular differentiation is governed by what Waddington (1957) called the *epigenetic landscape*, a complex panorama of networks and feed-forward loops that determine whether or not stem cells go into a lineage. As a developmental process, epigenesis is highly context-specific, following a chronological and topological logic. Especially during the embryonic phase for the normal development and further specificity of tissues and organ functions, certain genomic materials must be in the right place at the right time for the cell to assemble in a three-dimensional way with its neighbours. Throughout the life course, epigenetic processes remain sensitive to the complex biochemical processes in the cell and to its interactions with other cells, at the tissue and organ level, as well as to the impact of the endocrine system on other physiological systems. Variations in expression or silencing and in regulation do not entail changes in DNA sequences, either nuclear or mitochondrial. Phenotypic differences in local biologies are mainly due to the presence or absence of a resource at a specific time and place in a network of biochemical interactions. Compared with the genome, the epigenome – the epigenetic state of a cell at a particular moment – is highly dynamic and in constant flux.

On and off genomic expressions may occur very early during the embryonic stage, being 'generatively entrenched' (Wimsatt and

Schank 1986), directing and orienting further developmental and epigenetic paths. This has been called *epigenetic (re)programming*. Changes that promote the accessibility or closure of DNA to transcript factors are like tags, called *epigenetic marks*. Present in either parent's gamete – the sexual cell – or in both, they may be incorporated into the zygote, and maintained during mitosis and further cellular divisions of the egg and cell specialization. This has been called *epigenetic inheritance*, the molecular and biological mechanisms and organization of which seem to be very complex (Goldberg *et al.* 2007). Epigenetic marks can also be produced *ex novo* at any time, depending on biochemical events relating to embryonic, fetal, child, adolescent or adult experiences.[10] Many diseases can be explained by complex epigenetic processes at different moments and periods of life. More than 90% of breast cancers 'are not associated with any germline mutation' (Moss 2001: 89) but with epigenetic changes. Most cancers of any type are related to the aberrant methylation of gene promoter regions (Esteller *et al.* 2001, Lin and Maher 2010). Metabolic and cardiovascular problems (CVS), inflammatory bowel, auto-immune diseases, schizophrenia and autism spectrum disorders (ASD), among others, are also strongly correlated with genetic silencing or expression during the patient's life history.

As we shall see below, epigenetic processes are sensitive to personal embodied experiences which produce and are produced by somatic biochemical responsiveness to diverse environments. The case of monozygotic twins is exemplary, as it has been frequently selected as proof of genetic determinism and its narrative. Very similar genetically (although not necessarily identical, see Charney 2012) when born, they begin to develop phenotypic differences even in similar rearing environments; as they grow older these disparities increase. This happens mainly because of differences in DNA methylation and histone acetylation, as biochemical responses to diet and nutrition, exercise, restorative practices, emotional experiences, life hardships, stress and so on, all of which depend in part on income, social status and other factors (Fraga *et al.* 2005).

All things considered, it seems about time to take epigenesis and epigenetics seriously (Jablonka and Lamb 2002, Pembrey 2002, Ho and Burggren 2010). With regard to development and phenotypic plasticity, what matters are the patterns of gene expression (Fox Keller 2005: 4). It is

[10] Germline inherited epigenetic marks for disease propensity can be reversed: as in the saying: 'if any of your parents was hypertensive, run at least half an hour a day' (Shook *et al.* 2012). See also note 18, p. 82.

true that evolutionary change and differences between species are partly due to changes in protein-coding sequences; however they are also caused by modifications that affect gene regulation. The phenotypic plasticity that epigenetic changes can bring about explains part of the wide range of biodiversity that we encounter without our necessarily having to refer to genetic mutations. The same is true in human populations. There is evidence that height has the same endocrine aetiology for all humans. African 'Pygmies' (so-called), however, are much shorter than other people. Because of a defective response to growth hormones during puberty, due to very low values of baseline serum GH-binding protein, these populations do not experience the growth spurt that is usual in this period of life. It seems that both growth hormones and growth hormone receptors are under-expressed due to complex epigenetic regulations (Bozzola *et al.* 2009).

Epigenetic processes introduce another domain of variance for natural selection to act upon. In times of stress, error-prone DNA replication and activation produce a wide range of somatic variation. Epigenesis seems to tune this individual variability to that of the environment and its pressures (Rando and Verstrepen 2007). Epigenesis is not a random process as genetic mutation is assumed to be, but a stochastic phenomenon, probable but never certain (Waddington 1942). When environmental conditions change, cell physiology may change very quickly as well. When adaptive changes are long term, they may be fixed in the DNA. But when changes are produced on an intermediate evolutionary timescale, of between two and one hundred generations, cell memory and epigenetic inheritance contribute to their keeping in step with changes in the environment (Jablonka and Lamb 2005). The evolutionary consequences of this flexibility cannot be underestimated. Through *novelty*, *innovation* and *origination* (Müller and Newman 2005: 490ff), phenotypic plasticity due to epigenetic changes may be selected in a population by natural selection (West-Eberhard 2003, Pigliucci *et al.* 2006).[11] Midterm evolutionary adaptability could therefore be understood as phenotypic flexibility that is epigenetically inherited and environmentally directed, without having to wait for the natural selection of randomly

[11] There is evidence for the evolutionary role of DNA methylation (especially cytosine methylation) in animal genomes. It has been claimed that as much as 25% of the mutations that separate chimpanzees from humans occur at CpG sites, of which humans have only one fifth of the expected proportion (Elango and Yi 2008).

produced mutations, which induces adaptations that have a longer evolutionary span.

MATERIAL-SYMBOLIC AND POLITICAL WORLDS

One of the main claims of this chapter is that our anthropogenic built environments may induce certain epigenomic states and changes in our local biologies and at the population level, through the recurrence, institutionalization and modification of our material-symbolic practices. If genomic expression and silencing depend on epigenetic processes, and if epigenesis is environmentally sensitive, then our anthropogenic environments and the experiences they may individually or collectively induce have to be taken into account. This phenomenon has been called *environmental inheritance*. Strictly speaking, however, it is not so much environments that are inherited as the material-symbolic worlds in which we become the kinds of living creatures we are, hominized and humanized organisms – or in a word, *persons* – who will develop and grow as part of a new generation.

Gene-centred biology and its related disciplines do not consider the organism (or any other unit) as a co-building agent of its surroundings but rather as a passive recipient of evolutionary forces and environmental pressures dealt out by the natural selection of random mutations. No evolutionary changes are to be expected from organismic activity (Simpson 1953). However life systems, even in their simplest form, are distinctive because of their constructive eco-organization of a relevant scenario, a generative field of forms and relations, a landscape of variables affecting their own sustainability and that of their progeny (Maturana and Varela 1992). The concept of *niche construction* captures this complex autopoietic process of action in evolution (Bateson 1988). First proposed by Lewontin (1982) and later developed by Odling-Smee (1988) and others (Day *et al.* 2003, Odling-Smee *et al.* 2003), the theory of niche construction claims that organisms and environment are mutually constituted, including negative and positive feedbacks and possible co-ontogenies and co-evolutions.[12] Possibilities and constraints (*not* rules) for an organism's development, growth, adaptability and evolvability depend on the history of its structural coupling with other living and non-living systems with which it co-develops and co-evolves

[12] As omnivorous primates, our guts host millions of helpful bacteria with which we have coevolved (Ley *et al.* 2008). In that regard, we may call ourselves *metagenomic* organisms.

(Lythgoe 1979). As particular forms of existing and being alive, organisms are both the object and the subject of evolution (Lewontin 1983) and of development.

Organisms are particular places for growing and developing along the specific ways of a relational world. These ways unfold in and through their own life activities and are enfolded into their own specific morphologies, capacities of movement, modes of responsiveness and epistemic and emotional awareness (Ingold 2004: 216). Engineering their environments through niche construction, organisms become agents of change for themselves and for others. The complexity of humans' socio-material inter-relations among themselves, and with other species, artefacts, landscapes and inanimate objects, has had an unprecedented impact on other organisms' niches, environmental resources and climate. The political, economic, demographic and social transformations that occurred during the Neolithic era appear to have set in train a never-ending human appropriation of all kinds of niches at the cost of other organisms' ways of life. Through continuous processes, both determinate and contingent, of action-in-relationality and relation-in-activity, along with objectification and institutionalization, carried on through the generations, historically and culturally situated humans have built material-symbolic somatic and extra-somatic structures that can be stabilized, as scaffolding foundations (Shore 1996), to enable other evolutionary and socio-ecological structurings to emerge in particular times and places (Fuentes 2009a).

For any active human eco-organization, niche complexity is manifold, because human actions and practices are necessarily *material-symbolic*. This means that for us, there are no material actions void of a symbolic, ideographic and discursive quality by which they are made possible by and for an epistemic community of practice. By the same token, symbolic structures, rituals, imageries, discourses and the like are not empty vessels to be filled by a diversity of cultures. Rather, they exist only as long as they are practically produced in a particular relational world. For us, there is no ontological division between the material and the symbolic, between nature and culture.

For Marx, writing in the mid-nineteenth century, the anthropogenesis of our socio-symbolic-material conditions of existence held the key to human history. In his *Eighteenth Brumaire* of 1869, Marx stressed how the specific doings and makings of our forefathers (and mothers) reach the next generation, providing the latter with the contexts in which they can go on to develop and live their lives. We grow up, as

Marx realized, in environments built by the activity of our ancestors, who have shaped the scenarios available to us to make our own ways, just as we will do for our offspring and their successors (Ingold 2004:217). We can therefore speak of our ecological niche as one that establishes the possibility of our human and humanized existence as active recipients of our predecessors' practices, representations and experiences – a niche that we will inevitably change by way of our presence and activities, as will our children and grandchildren in carrying on their own existence.

Stotz (2008) speaks of *developmental niche construction* as the provision of a set of stable environmental resources by the organism's species antecedents, its 'parents', as well as by its own life activity, allowing it to develop in a reliable and predictable pattern of existence. West and King (2008: 384) call this 'exogenic inheritance', the continuation over generations of a reliable and stable environment that can nonetheless 'adapt rapidly to changing ecological circumstances'. The concept of *ontogenetic niche* refers to how species-typical behaviours develop through active, context-dependent processes within a complex genomic and epigenomic regulatory system that includes molecular, cellular, ecological, social and political experience, together with other developmental resources such as social, cognitive and emotional stimulation in the pragmatic caring world of parents, siblings and others who provide 'nurturing' and affordances for development (Alberts 2008a). The recurrence of the same or similar environmental possibilities and constraints in the form of normally occurring sensory stimulation has a bearing on genetic and epigenetic activity. Similar, species-specific environmental landscapes probabilistically induce similar developmental systems in organisms (Gottlieb 1971).[13]

Material-symbolic practices of parenting and care-giving in familial environments are critical for nutrition and protection, and for providing the sociocultural knowledge that enables children to achieve active and balanced biopsychosocial growth and maturity. As both psychobiology and the anthropology of child-rearing and socialization have shown, parental care is one of the most powerful forces in stabilizing the parameters of the ontogenetic niche. Nurturing activities, control of alertness and sleep, attention and posture, body contact and manipulation, perceptual and sensorial stimulation, social exposure

[13] For example, the capacity for copulation in male rats is not innate but acquired in the most probable environment that the developing rat system encounters (Ariew 2006). See also Avital and Jablonka (2000).

and its contexts, attention guidance, the formatting of communicative skills, direct and indirect speech, social referencing, and so on, are co-produced as material-symbolic practices in communicative socio-cultural settings, in which the infant or child participates either actively or vicariously (Ramirez Goicoechea 2009). Indeed if parents and caregivers make children, it is equally clear that children make parents or caregivers, in a non-isomorphic reciprocal process. Among much other evidence for this, it has been reported that bonding with the child lowers testosterone levels in the male parent.

Care, parenting styles, stimulation and the home environment all have a major impact in mediating children's early mental development, even in spite of low economic status (Kim-Cohen *et al.* 2004). The impact of the quality of the home environment on the child's behaviour is greater when coupled with variations in socio-economic status and the differential access to experiences and resources that goes with them, showing that it is the combined effects of all these variables that best explains differences in children's development (Bradley and Corwyn 2002, Propper and Rigg 2007). As adoption studies show, genetic predispositions for some psychiatric disorders (such as schizophrenia, autism spectrum disorder and attention deficit hyperactivity disorder) need the triggering force of the environment for their expression. Thus, genetic factors are evoked in the child's phenotype depending on the quality of the family caring environment (Maccoby 2000) and on the child's active responsiveness.

A relationship between socioeconomic conditions and glucocorticoid stress response in childhood has been established in an ethnographic study in the Dominican Republic (Flinn and England 1997). It was discovered that this influence was indirect, affecting children through the quality and intensity of child care. Release of cortisol may help to modulate perceived psychosocial and cognitive-emotional affronts, at the cost of immunity and other health shields. Difficult family environments such as those involving partner disruptions, caregivers with economic and emotional problems, traumatic events, and so on, affect the quality of care, with specific consequences for progeny, who may go on to develop health problems. This is also connected with individual factors such as the child's vulnerability and resilience, depending on age, sex and personality (Elder 1974).

Childhood sexual abuse also sets off disturbances in many interconnected systems and subsystems of the body, potentially leading to a plethora of physical and mental symptoms and disorders. It may coincide with family violence, separation or divorce, substance abuse, mental illness, or crime, the intersection of which may produce a cascade of interwoven effects (Maccoby 2000, Fullilove 2009). In a study of adults with psychiatric illness, the authors concluded that the impact of childhood sexual abuse on the illness burden was 'roughly comparable to the effects of adding 8 years of age'. Its impact on daily activities and body pain was even higher, as if adding twenty years of age to the individual (Talbot *et al.* 2009).

CROSS-GENERATIONAL EPIGENETIC EFFECTS

There is a growing body of evidence from animal models and longitudinal surveys of human families that environmental factors have complex ways of influencing the epigenetics of human biology throughout the life course, and especially during specific developmental stages. It is not that our socially built environments directly trigger gene expression or silencing. Instead, they constitute biosocial settings for particular developmental and growth processes through biochemical and hormonal interactions. Somatic and extrasomatic environments may induce epigenetic changes through biochemical responsiveness. It is through these chemical and hormonal changes that epigenesis takes place at the molecular, cell, tissue and organ levels.

What have been called *mother epigenetic effects* during periconceptional, fetal and early infancy periods have a strong influence on the child's phenotype and health. The uterine and perinatal experiences of the fetus and baby take place in developmental settings[14] in which the mother's biochemical and hormonal situations may induce specific epigenomic states in progeny (Gluckman *et al.* 2008). These maternal effects of epigenetic transmission may have a huge impact on the baby's development and on its subsequent health during childhood and adulthood. Some of these early embryonic and fetally induced states may become deeply embedded in early ontogeny, working as stabilized structural enhancers or constraints on further developmental processes.

Biochemical, hormonal and sensory-motor psychosocioemotional interactions in mammals between mother and fetus and baby

[14] 'It is an inconvenient truth that the onset of behavior is before birth' (Alberts 2008b: 270). In prenatal developing systems of experience (Gottlieb 1997), 'each sensory system begins to function while still undergoing maturation . . . so each system could contribute to its own normal prenatal (as well as postnatal) development' (Gottlieb 2001: 44–5).

take place especially through placental and umbilical cord exchanges, breastfeeding, and through early stimulation, control and regulatory behaviour. For example, studies of the Norway rat (*Rattus norvegicus*) show that suckling is not innate but the result of olfactory stimuli presented by the mother's body and hair. The mother shapes the little rat's behaviour, in responding to these stimuli, into an exploratory search which ends up with its finding the teat and then suckling (Alberts 2008a). Sexual differences in gene expression in the developing spinal cord of rats are due to the mother's differential liking of pups' genitalia which, in turn, results from the greater expression of chemicals in male pups than in female ones (Moore 1992). Through intrauterine influences on development, offspring of overweight and diabetic mothers have a strong possibility of suffering from insulin resistance during childhood and adulthood (Yajnik 2004). Mice with a genetic propensity to develop cardiovascular diseases could have this propensity – and therefore gene expression – reversed by changing their mothers' diet when breastfeeding (Mcleod *et al.* 2007). In humans, too, the origins of some cardiovascular diseases in fetal development have been explained by epigenetic mechanisms under the label of the *Barker effect* (Barker 1995, 2004).

While providing close body contact and the stimulating effect of the mother's odour and heat, all of which are important for biopsychological and emotional balance, breastfeeding also bolsters children's immunity. Early development of gut flora through breastfeeding protects the baby from further allergies, diarrhoea and inflammatory disorders, as well as from type 1 diabetes. Indulgent lactating mothers reduce the anxiety and reactivity of offspring in response to stress, an effect that can be carried over to the third generation. The increase of non-medically justified caesarean births has led to a reduction in babies' contact with mothers' vaginal flora, which is important for building up their own immune systems. Caesarean section increases the risk of type 1 diabetes by 20% (Cardwell *et al.* 2008), and likewise for the risk of asthma (Thavagnanam *et al.* 2008). The physiological and psychological consequences of assisted reproduction technologies (ART) and in vitro fertilization are also under scrutiny for their possible epigenetic consequences. The manipulation of sperm and ovocytes in ART, especially if there are multiple births, is associated with an increase in the risk of several birth defects such as cleft lip, cleft palate, gastrointestinal and other problems (Reefhuis *et al.* 2009). It has been proposed that another risk associated with in vitro fertilization may come from epigenetic changes brought about by the methionine content of proprietary commercial media used in ART that may be linked to children with Beckwith–Wiedemann syndrome, Angelman syndrome and retinoblastoma (Niemitz and Feinberg 2004). Assisted reproduction technologies may also be linked to some cases of attention deficit hyperactive disorder (ADHD) (Ana Vidal, personal communication, 2004). Although less known, paternal effects are also relevant. The father's presence and his involvement in caring practices have a strong influence on the maturation of offspring in some multi-male primate societies (Charpentier *et al.* 2008). Analysing data from the ALSPAC Study,[15] it has been reported that fathers who had started smoking before the age of 11 had sons with higher body mass indices; this did not happen with daughters (Pembrey *et al.* 2006).

Practices, representations and beliefs concerning the social and political distribution of health, diet and nutrition, gender and sexuality, hierarchy and autonomy in decision-making, work conditions, stress, trauma, exposure to pollution and toxins, and so on, have consequences not only for phenotypic plasticity at the somatic level. Our forefathers and foremothers provide us not just with specific, anthropogenic and extra-somatic environments for playing out our lives. Besides an environmental/ontogenetic niche, they also bequeath a genomic/epigenomic legacy. Parents' and ancestors' epigenomes may be incorporated in their offsprings' gametes and then retained in the egg cell, in meiosis and in the cell's further divisions. Incorporated in the germline – although not in the DNA – these epigenetic marks may have important health consequences for the new generation, affecting glucose metabolism, fat deposition, blood pressure, neurophysiological response to stress, cognitive development, food sensitivity, and susceptibility to allergies, inflammatory bowel, cancer and many other conditions. The molecular mechanisms of these cross-generational phenomena are still under investigation. I present here some empirical cases, from studies extending over three generations.

Bygren, Kaati and Edvinsson (Bygren *et al.* 2001) studied harvest and food prices, demography and health changes in a small and isolated northern Swedish town called Överkalix. They were able to

[15] ALSPAC (The Avon Longitudinal Study of Parents and Children, formerly the Avon Longitudinal Study of Pregnancy and Childhood) is designed to understand the effect of environmental pressures on the individual's genotype, and how they combine to influence health and development. By 2010 data from nearly 10 000 children and their parents had been collected, from early pregnancy until children were 8/9 years old. See http://www.bristol.ac.uk/alspac/.

establish a strong correlation between ancestral food supply and the longevity of the third generation. They also showed a link between food availability to ancestors and the cardiovascular and diabetic risk to descendants (Kaati *et al.* 2002). They discovered that the correlation between food access and reduction of longevity and the increase in cardiovascular and metabolic risk only held when it was the paternal grandfathers who had access to the food supply during the slow growth period, established for boys between 9 and 12 years of age (Bogin 1999), before the pre-pubertal peak, when less nutrition is needed. Quite to the contrary, the grandchildren of grandparents who experienced famine during the same period were less prone to suffer from cardiovascular conditions than those of ancestors who did better during those years. Reanalysing the Överkalix data, Pembrey *et al.* (2006) discovered that paternal grandmothers' food availability was also associated with gran-daughters' mortality risk, but not with that of grandsons. Some of their conclusions were: (1) that a sex-specific male-line transmission is linked to chromosome Y in humans; (2) that the slow growth period is crucial in epigenetic marking, especially in gametogenesis, which in males occurs during adolescence, and in girls mainly during fetal life; (3) that the biological system is capable of short-term responses to rapidly changing environments; and (4) that ancestors' environmental exposure and bio-cultural responses have transgenerational effects. Those epigenetic changes that also appeared in individuals of the third generation did not affect their DNA, nuclear or mitochondrial, nor did these individuals have to undergo the same or similar experiences to their grandparents, in order for this epigenomic coincidence to be exhibited.

Another interesting case concerns the impact of famine on pregnant women and their future babies during war periods. Mothers' malnutrition can have a severe impact on their babies' weight, besides leading to other health problems such as allergies and asthma. Studies of pregnant mothers who suffered from the Dutch Famine during the last years of the War showed that their children were underweight; their granddaughters also suffered reduced fetal growth, even though their mothers had not experienced the severely restricted diets of their grandmothers (Susser and Stein 1994). It seems likely that epigenetic markings explain these cross-generational effects (Lumey 1992, Roseboom *et al.* 2001). Similarly lasting effects of famine due to war conditions have been reported for other European countries as well (Van Den Berg *et al.* 2011). Trauma, violence and abuse should also be taken into account. The placental environment protects the fetus up to a certain point from the cortisol produced by their pregnant mothers when suffering from stress or anxiety. But maternal cortisol production during pregnancy after severe traumatic stress leads to high levels of cortisol in the second and even third generations. This alteration of stress reactivity has been reported in pregnant mothers who survived the Holocaust and in their grandchildren. Similarly, offspring of gestating mothers exposed to the 9/11 attack in Manhattan gave birth to babies with hypothalamic-pituitary-adrenal axis alterations as well as stress response disorders in early childhood (Yehuda and Bierer 2008). It has been suggested that many of these uterine experiences induce specific types of responsiveness due to the triggering of epigenetic changes in the fetus.

RACIALIZED BECOMINGS

Different forms of power, socioeconomic organization and classificatory systems produce specific experiential environments for individuals and groups. The disenfranchisement of large parts of the population (Lende 2008) increases their exposure to vulnerability and health risk.

Endocrinologist Robert Sapolsky (2004) states that status has a strong influence on quality of life in social species that are organized into ranked hierarchies. In humans, economic deprivation, malnutrition, poor housing conditions, exposure to pollution and toxins, lack of education, poor access to medical facilities, diseases, epidemics, along with catastrophes, discrimination, exploitation, violence, abuse and suffering, have direct consequences for life expectancy and health.[16] Low social status has been correlated with psychoemotional stress and cell metabolism and oxidation (ageing), reducing life expectancy through telomere shortening after successive chromosome replication (Cherkas *et al.* 2006). Depression of the immune system can also lead to cardiovascular suffering and possible changes to the reproductive cycle.

Occupational and work-related diseases and deformities are well-known in historical demography. There are strong links between bones, culture and activity (Fausto-Sterling 2005, 2008). Ethnic devaluation, income and health are strongly connected in some populations (Walls and Williams 2004). Many global infectious diseases follow the same distribution as malnutrition and poverty. Indeed there is no doubt

[16] Conversely leisure activities, as part of a (socioculturally and historically defined) pleasant life, influence endorphine production, contributing to well-being and stronger self-defence against infections, which, in turn, increase longevity (Hyyppä *et al.* 2006).

that we can speak of a political economy of health, suffering and risk exposure (Goodman *et al.* 2005). *Globesity*, a term recently introduced by the World Health Organization, concerns the global epidemics that are becoming a national health crisis in the USA (Ford *et al.* 2002), as in many other developed countries and emergent economies. Obesity is part of the metabolic syndrome which also includes insulin resistance and hypertension, especially among African American women, though not among West Indian African-Americans, male or female, or Africans (Cooper *et al.* 1997). It is also a precursor of stroke and coronary heart disease, and hence of premature death associated with these conditions. In the USA, statistics show that African Americans suffer twice as much from this syndrome as white Americans. Mortality due to these health problems is much higher in the first group than in the second (Nesbitt and Victor 2004, Mensah *et al.* 2005).

Some epidemiologists have tried to establish a causal connection between the prevalence of this syndrome and hereditary genetic factors. In 2005, the American Food and Drug Administration approved the first racially specific pharmaceutical treatment for heart failure among African Americans, opening up a new market for pharma-industries interested in a new possibility for medical customization and commodification: racial difference. The geneticization of race in medical research is becoming a serious challenge both to science and to non-discriminatory policies and ideologies (Palsson 2007b). Actual genetic diversity has nothing to do with racial categorizations (Armelagos and Goodman 1998), which are politically and historically constructed, essentialist classifications based on irrelevant phenotypic differences and adduced for purposes of discrimination and segregation. Though clusters of genetic similarity can be found in some populations, they correspond not to racial partitioning but to probable geographical areas of origin (Templeton 1998). Race cannot be a legitimate scientific explanation for any genetic variability, or for actual health disparities (Goodman 2000).

If race is not the reason for the differences in susceptibility to the metabolic syndrome between African Americans and white Americans, the explanation must be sought elsewhere. Although hypercaloric diets, sedentary life and differences in brown adipose tissue that transforms food energy into heat[17] are important factors in obesity, they cannot

[17] Abdominal adipocytes are big white fat cells, the main producers of cytokines, which are directly related to the endocanabinoid system, the deposition of lipids in the arteries and inflammatory processes (López-Jaramillo *et al.* 2005).

explain why people with similar lifestyles show such a disparity. Neither prematurity nor body weight when born are genetic. One of the most relevant consequences of nutritional stress in gestating mothers, together with emotional stress during pregnancy, is that they give birth to premature or underweight babies. Nutrition and diet are important factors in the developmental environments for all mammals. Studies have yielded evidence in many species for a link between malnutrition, lack of necessary proteins and nutrients, and placental size, leading to lifelong changes in the body and its systems, including mental and psychological disorders in later life (Antonow-Schlorke *et al.* 2011). Loss of the caring environment of the placenta due to premature birth (Bainbridge 2003) has been correlated with organic and functional underdevelopment in fetal brain and lung tissue, plus the high probability of medicalization in the baby's and infant's early ontogeny, if not throughout the course of life.

Pre-term or low-weight babies exposed later in life to rich caloric diets – such as cheap junk food – are not biologically prepared for such a quick metabolic change. They tend to show a higher propensity to weight gain, and metabolic glucose and lipid disorders, which are the precursors of later cardiovascular diseases. It is as if they were already able to make the most of their diet in a context of food shortage when born – as if their metabolic threshold for coping with a caloric diet without major health consequences would be lower than that of full-term and normal-weight babies. This mismatch between pre- and perinatal environments and those of childhood, youth and adulthood may explain the prevalence of future obesity and its correlated diseases.

Let us return to the aforementioned health differences in the USA. African American mothers tend to give birth to underweight babies, either pre- or full-term, twice as often as white American mothers. But this is not the case for African-born mothers giving birth in the USA, who tend to have full-term normal-weight babies. By contrast, the African American-born daughters of African-born mothers, entering a similar niche of social exclusion during pregnancy, will tend to deliver premature or low birthweight babies (Kuzawa and Sweet 2009). The prevalent environments for many African American pregnant mothers are characterized by malnutrition, unhealthy lifestyles and precarious living conditions, limited access to prenatal and postnatal medical care, and strong psycho-emotional stress (depression, anxiety, overwork, single motherhood). It is the influence of the mother's specific and embodied responsiveness to environmental distress and hardship, linked to the effects of social exclusion and discrimination practices within a racist political economy, along with her personal psychoemotional experiences and home atmosphere, that produce developmental niches for children which may increase their future risk of suffering from the metabolic syndrome, and that better explain the health differences between African American groups and Americans of European origin.

We know that the embodied responsiveness of the embryo and fetus to specific uterine environments, in the form of epigenetic changes, can be imprinted in the child's gamete, and passed later to his or her children, and through them, to future generations. Kuzawa and Sweet (2009) suggest that the chronic effects of social-environmental exposures, if epigenetic conditions are maintained, are passed to the following generations and even to their offspring. Developing health problems jeopardize social and economic opportunities both at the personal and at the group level, triggering repetitive self-induced life cycles of poor health and social exclusion. This ontogenetic loop may be re-enacted in the offspring's local biology, and passed to their future generations through the germline. It is the complex stochastic embodiment of multiple interactive systems of *life-in-the-making*, of socio-political, economic and ideological discriminatory structures and practices, together with individual and biographical experiences, that produce certain kinds of maternal physiological and psychological responsiveness. These constitute the child's pre-, peri- and postnatal environments which produce and maintain specific epigenetic states that may, in turn, be passed as epigenetic marks to future generations. This is why we can speak of matrilineal family cycles of biosocial suffering and exclusion in racialized groups. These cross-generational maternal effects may contribute to the perpetuation of social structural cleavages and social exclusion. Individual life courses or unhealthy lifestyles that depend on income and social status, but also on socialization in group traditions and values, can reinforce these matrilineal epigenetic phylogenies of biosocial becomings.

CONCLUSIONS

We began this chapter with a critical review of the gene-centred view of human phenotype and behaviour. We demonstrated the relevance of epigenetic processes in the context of an epistemological shift towards recognizing the role of developmental systems in the construction of life and phenotypes. We went on to emphasize the importance of human material-symbolic productions in the construction of developmental niches. We attempted to correlate these niches with differences in health and with a political economy of inequality and deprivation in some populations. We explained the cross-generational effects of epigenetic changes, both in ontogeny and through the germline. Within a biopsycho-sociocultural and political framework, we argued for an explanation of human becomings in terms of the material and symbolic legacy of racism rather than genetic inheritance. We showed how practices, representations and discourses transcend generations, and how social doings and makings are bioculturally and politically reproduced in cross-generational cycles of social exclusion and discrimination. In so doing, we have attempted to combine an emphasis on social exclusion and ethnicity with a fully interdisciplinary approach to human affairs. Some of the conclusions we may draw from this discussion are the following:

- Epigenesis is key to the post-genomic shift from genetic determinism to a focus on the interactional networks of stochastic genomic processes in environments of development. This shift is fundamental to a biopsycho-sociocultural approach to human becomings and to showing how life is *in-the-making*.
- Political, economic, ideological, historical and biographical factors are embodied by pregnant mothers, influencing the conditions for their chidren's development, which in turn leads to specific epigenetic states with probable health consequences throughout the life course, impinging on people's socioeconomic careers and status.
- Parental and ancestral epigenetic marks may transcend the generational cohort in which they first appeared, crossing over to following and future generations and establishing a propensity to suffer from certain health conditions.

If the genome is not our destiny, nor does it lie in our epigenome. Although epigenetic marks can be passed on without experiencing the life conditions that once triggered them, particular attention may be paid to reversing their effects. We are the repositories of our genomic and epigenomic legacies not only for ourselves, but also for our descendants and their future progeny. This involves ethical and political dimensions of how we decide – or are allowed – to carry on our present lives. The responsibility is not only personal but also collective. Proper interventions during the life course, especially in crucial human developmental stages such as those of the embryo, fetus, infant, child and adolescent, can counteract the negative biological effects of embodied political economies of inequality in human becomings. Public policies should focus on socioeconomically and politically induced lifestyles that produce biochemical, hormonal and neuro-psychological predispositions to unequal human development. Transformations and improvements of developmental niches and long-life environments for young people, adults and the elderly are possible, provided that a political commitment exists and the necessary resources are allocated.[18] Last but not least, the political economy, historicity and socio-cultural dimensions of human epigenetic landscapes, and their possible cross-generational perpetuation, open up a promising scenario for interdisciplinary collaboration between the life sciences and the social sciences, contributing to a much needed eradication of dualistic thinking.

ACKNOWLEDGEMENTS

This chapter has been made possible by a period of sabbatical leave (2009–2010) from my university (UNED, Madrid, Spain) that I enjoyed at the Human Systematics Laboratory of the Universitat de les Isles Balears (UIB, Majorca), thanks to the academic generosity of Professor Camilo Cela Conde. I greatly appreciate Dr Nancy Konvalinka's help in taking over some of my academic duties during that time; also for her kind revisions to my written English. I am also grateful to Daniel Lende for useful comments on an early draft, which was presented at the EASA 2010 Conference at Maynooth (Ireland). I thank Tim Ingold and Gisli Palsson for organizing such a fruitful workshop. This paper is dedicated to the

[18] The effects of unbalanced nutrition and health can be reversed. With a three-month Mediterranean diet based on cereals, fish, olive oil and some wine, these markers change drastically (Estruch *et al.* 2006); also, prostate cancer patients who showed intolerance to chemotherapy could control oncogene expression by means of a hypocaloric diet, one and a half hours of exercise each day and daily meditation (Barnard *et al.* 2008). See also *The Southampton Initiative for Health* (SIH), a clinical and social intervention to improve the health lifestyles of women and their offspring in Southampton, UK, at http://www.mrc.soton.ac.uk/index.asp?page=291.

Biosocial Becomings team which was born on that occasion. Although references and notes have been reduced to a minimum, I apologize for the overwhelming number of citations. These are intended to provide a tentative map of possible itineraries for those interested in exploring the issues in more depth.

AGLAIA CHATJOULI

5

Thalassaemic lives as stories of becoming: mediated biologies and genetic (un)certainties

My ethnographic journey of exploration into thalassaemic lives took place in an era characterized by the catalytic role of biomedicine in widening the spectrum of theories, tools and concepts available for understanding, constructing and reconstructing life. Bodies, biologies, organisms, genes, genomes, tissues, organs, behaviours, intentions, environments, human and non-human animals, plants as well as inanimate objects become increasingly problematic when fixed as concepts and entities. We are witnessing the inevitable conflation of nature and culture, biology and society, body and mind, of underlying dualisms shaping previously well-separated disciplines focusing on the study of human life. Even though our increasing knowledge of the potentiality of human and non-human genetic material and how it interacts with many other forces of life has led to persisting determinisms and reductions, other theoretical currents have been equally informative, more holistic and inclusive in nature, offering insights into the ways genetic material is situated within – but is not a primary determinant of – the process of living. If anthropology has a tradition of fixing dualistic understandings of the world, it has also contributed to the systematic appreciation of diverse ways of living, emphasizing context sensitivity, self-understanding, agency, and relationality, all of which have tended to erode this fixity.

For this chapter, I have drawn from ethnographic research on thalassaemia in Greece during the period 2003–2006. Thalassaemia is a well known monogenic disease. It is broadly defined as an inherited disorder caused by defective and imbalanced globin production

Biosocial Becomings: Integrating Social and Biological Anthropology, eds T. Ingold and G. Palsson. Published by Cambridge University Press.

resulting in severe and often life-threatening anaemia. It belongs to a wider group of inherited human haemoglobin disorders (linked to structural variants and hereditary persistence of fetal haemoglobin). The term thalassaemia is derived from two Greek words: θάλασσα (thalassa) meaning the sea and αίμα (aima) meaning blood. The term itself reflects the high prevalence of the condition in Mediterranean countries. The β-thalassaemias present the biggest health problem globally when compared with the other types (α-, δβ-, εγδβ-), and represent the most common type in Greece.

My key research aim was to understand people with thalassaemia, otherwise known as thalassaemics.[19] Lay self-understandings of biology, of the person, and of life were at the centre of ethnographic attention. Fieldwork included visiting blood transfusion centres, attending patient group meetings, national and international Thalassaemia conferences and cultural events celebrating World Thalassaemia Day or promoting voluntary blood donation. I mostly met and conversed with people diagnosed with *β-thalassaemia major* and *intermedia*, not with people with *thalassaemia minor* also known in Greece as 'people having the stigma' (heterozygotes who usually do not manifest health problems or asymptomatic carriers). In other words, I met with patients needing life-long treatment, regular blood transfusions and iron chelation. In severe cases, no treatment may lead to early death. Various serious health complications occur in the case of thalassaemia major, especially when dealing with untreated or poorly transfused individuals (Galanello and Origa 2010). Such problems include growth retardation, pallor, jaundice, poor musculature, hepatosplenomegaly, leg ulcers and skeletal changes resulting from bone marrow expansion. In addition, many secondary problems arise due to transfusion therapy and the subsequent iron overload such as endocrine complications resulting in failure of sexual maturation and the development of diabetes mellitus, among others. Iron overload may also cause dilated myocardiopathy, liver fibrosis and cirrhosis. Patients with thalassaemia intermedia usually present later in life with moderate anaemia and may not require regular transfusions (Galanello and Origa 2010: 1).[20]

[19] In the context of this chapter, by thalassaemics I mean individuals who belong to different biomedical diagnostic subgroups. In this I follow the way the term is used by the people themselves. More specific diagnostic labels are discussed and used when needed, as was the case in the field.

[20] Intermedia cases are usually also characterized by osteoporosis due to 'hypertrophy or erythroid marrow with medullary and extramedullary

The aforementioned biomedical categories (major, intermedia and minor) reflect degrees of clinical severity and attempts to formulate meaningful genotype–phenotype relationships diagnostically, prognostically and therapeutically. Despite the remarkable progress in discovering many mutations in the beta globin gene, as well as a multitude of other genetic factors affecting the disease phenotype, categorization has been difficult due to the notable heterogeneity of this monogenic disease.

My personal interest in thalassaemia was born in part from ongoing anthropological discussions regarding the dynamic role of the biosciences in reconfiguring (human) nature. What would someone embodying a diagnosed genetic difference have to say about being born with a 'genetic anomaly' leading to chronic illness and often to physical differences or deviance, someone living in a biomedicalized environment where gene-talk has infiltrated the lay imaginary? The timing was right in many ways. During the last decades, biomedical progress has enabled thalassaemics in some parts of the world actually to *keep on living*. Most of the 3000 adult Greek thalassaemics may never have reached adulthood had they been born in a different time or place. These people have witnessed key changes related to therapy, survival and life-expectancy rates and quality of life. Furthermore, the emergence of disability activism and a growing discourse supportive of 'the right to difference' have further altered the intersubjective experience of social inclusion and exclusion.[21] In addition to treatment, prevention has been a key public health goal in many countries. The Greek National Prevention Programme for Thalassaemia, initiated in the 1980s, is a paradigmatic example of new eugenics aimed at all couples planning to have children and has been successful in nearly eliminating new thalassaemic births. It seems therefore that the adult thalassaemic population in Greece has reached its peak. At least in this part of the world, the social institution of prevention has successfully controlled the problem, thus making the current generation the last one. This eventuality alone, depicted in the transitory making of a

haematopoiesis and problems with the spleen, liver, lymph nodes, chest and spine due to masses of erythroid tissue. Bone deformities, gallstones, painful leg ulcers and increased predisposition to thrombosis are additional complications' (Galanello and Origa 2010: 1).

[21] In the world of thalassaemia, a common identity has emerged, linked to the pathology, informed by advances in treatment and prevention, and moulded in the course of three decades of collective activism reflecting in many ways aspects of genetic citizenship (Chatjouli 2012).

thalassaemic population, highlights aspects of the relational and biosocial nature of human lives.

Examining thalassaemia in a place like Greece yields insights which resonate with the programmatic intentions of this volume. Firstly, it can be used to argue against a culture/nature dichotomy by demonstrating the lived-embodied normality and naturality of biomedically mediated lives. In the thalassaemic context, hybrid natural-technical or relational biologies, along with assisted-enabled-mediated bodies, delineate the 'natural' and real way of being. This state is not contested or sensed in dichotomous ways. Secondly, given that thalassaemia is among the most-studied of monogenic diseases, it is an excellent example by which to trace the spread and normalization of genetic reasoning. The linking of genetic difference to diseased phenotypes has made carrier couples an uncontested target for population screening and prevention programmes. In the context of geneticized biomedicine, reductionist reasoning is informing the lives of born or unborn thalassaemics, carrier couples, treatment protocols, medical and gene therapy research and healthcare and welfare biopolicies, highlighting how a specific conceptualization of biology constructs the relational and biosocial process of human becomings. By focusing on the lives of thalassaemic women and men and not solely on their biomedically defined natures, the wider complexity of living can be revealed, or at least parts of it, pointing to the multitude of interactants informing the development of the person, as well as the limitations of gene centrism.

For such a reading of thalassaemic lives, I have followed lines of analytical thought that point to the merging of nature and culture, of the natural and the artificial-technical, and of biology and society. Of equal importance are works that emphasize many other local and global notions, besides those comprising the biomedical conceptual toolkit, which are equally important in understanding and constructing life (Haraway 1985, Strathern 1992a, 1992b, Descola and Palsson 1996, Cambrosio, Young and Lock 2000, Franklin and Lock 2003, Goodman, Heath and Lindee 2003, Ingold 2004, Landecker 2007, Rose 2007, Palsson 2008). The shifts and changes brought about by advances in biomedicine highlight the artificiality of nature together with the intentional involvement of human action. Attention is drawn to the co-production of nature and culture (Jasanoff 2004) and to the interactions among nature, science and society (Latour 1993). Pronouncements on nature that once brought a sense of certitude and wisdom are consistently undermined (Haraway 1985: 41).

An analytics of biosociality (Rabinow 1996) enables us to follow the ongoing transformations of the category of life and the conceptual changes brought by the new genetics. The dangers inherent in using biology as a metaphor for society become increasingly apparent following the realizations that 'nature will be moulded on culture understood as practice' and that 'nature will be known and remade through technique and will finally become artificial, just as culture becomes natural' (Rabinow 1996: 99). A biosocial approach to studying emerging biologies, normalities and socialities in diverse contexts enables us to better understand the ways in which new biotechnologies interact with other socio-cultural forces, informing people's relational lives and their involvement in the art (*techní*) of (making) nature. As Palsson notes, 'local understanding and interpretation of genetic analyses are likely to depend on a host of factors, including how people construct nationhood, citizenship, subjectivity and relatedness' (Palsson 2008: 558).

This chapter is furthermore inspired by theoretical projects that attempt to problematize and historicize biology. These projects have given careful attention to different ethnographic contexts (Palsson 2007a) and have stressed the unity of genes, organism and environment, adopting a critical stance against gene centrism especially in the light of advances in fields like epigenetics and developmental systems theory (Ingold 2001a, Oyama, Griffiths and Gray 2001, Palsson 2007a, Barnes and Dupré 2008). In this chapter, emphasis is therefore placed on the biosocial, processual, relational and developmental character of human becomings as well as on the embodiment of human activities. In light of such theorizing, I have attempted to illuminate thalassaemic lives in terms of *persons-organisms living in an environment*. As Ingold writes:

> The body is the organism, and the process of embodiment is one and the same as the development of that organism in its environment . . . Substituting life for mind, and organism for body, the notion of a mindful body may be replaced by that of living organism, a substitution that has the effect both of restoring human beings to their proper place within the continuum of organic life, and of laying the Cartesian dualism finally to rest. (Ingold 2001a: 259)

Humans, Ingold argues, are becomings 'whose identities are established, in the first place, not by received species- or culture-specific attributes but by relational accomplishment' (Ingold 2010c: 363).

Being born with certain identified genetic characteristics that often contribute to physical differences and constraints and that may lead to the death of the person-organism (depending on the environment) highlights the restrictions of materiality, and calls for modes of reasoning rather more inclusive than social constructionism. Reflecting on such characteristics (genetic, physical, biological) without undermining their positioning and embeddedness within complex biosocial relations – no matter how striking these characteristics may seem from a certain point of view and in comparison to a normalized majority – enables us to approach living-in-this-world without succumbing to dualisms. Instead, we can understand differences as emergent within contexts of ontogenic development. The emphasis is not on links of cause and effect but on the multitude of relational and processual engagements with the environment. 'The life cycle of an organism is developmentally constructed, not programmed or preformed. It comes into being through interactions between the organism and its surroundings as well as interactions within the organism' (Oyama, Griffiths and Gray 2001: 4). Following the developmental histories of thalassaemics, one finds lines, links and flows leading to significant information, relationships, cosmologies, drugs and policies. These lines and flows reveal the complexity of the person-organism as well as the implications of specific constructions of the biological person.

Diagnoses of clinical severity following specific readings of biology inform the construction of β-thalassaemia subcategories, treatment protocols and state benefits. Certain genetic differences and certain phenotypes become the focus of biopolitical attention, further influencing the developmental unfolding of life. Timely or delayed onset of blood transfusion and the quality of transfused blood are socially conditioned and fundamental to disease progression and survival rates. Endocrinological biomedical monitoring mediates sexual maturation and reproductive capacity. Treatments such as iron chelation halt organ deterioration. The formation of a collective thalassaemic identity and an active thalassaemic community (local and international) facilitate the exchange of knowledge and experience among patients, resulting in demands for better treatments, more research and state support, thus changing the course of thalassaemic lives in manifold ways.

BIOMEDICALLY MEDIATED YET NORMAL LIVES

They are freer nowadays (referring to her co-patients). The conditions are much better and that is good. Plus there are almost no new births,

> perhaps only a few. There is hope. The older generation, they are created, they have their circle, their lives, they have progressed. They haven't stayed in the same place. I may say, without sounding overoptimistic that thalassaemia is amongst the less serious cases that may happen to someone, and I am sure of that. If you deal with it that is, I mean if you deal with it medically. What we have to deal with is unfortunately a lot: transfusion, heart problems . . . But if you deal with it in the right medical way you have nothing to be afraid of. (Nandia)

> In general I don't feel it, that I have thalassaemia because I have a normal life, I go to the gym, I go here and there, I do this and that. Then okay, I have to be at home to hook onto the chelation device. But now I don't even have that because I use the pill [for iron chelation]. The device I may use only three times a week. But now I don't mind, I put in on myself, I don't feel anything. I don't mind it, it's like having to eat and having to go to bed. I had to get used to it when I was little, the piercing and all that. Sometimes it might hurt when you have it all night but now with the pill it is easier of course. (Georgia)

The health status of thalassaemic individuals varies according to many factors such as severity, heterogeneity, availability and compliance to treatments, state and family support, and diet – to name a few. Some thalassaemics may have had a better medical attendance when they were young, some may not have the patience or the facilities to follow the treatment protocols meticulously. Despite the differences, one thing is common, at least to the individuals who have presented with the problem from a very early age, and that is the lack of an embodied memory 'without thalassaemia' and 'without biomedical interventions'. Embodiment of transfusion, on the other hand, being the most prominent and life-long act of biomedical mediation (together perhaps with iron chelation), is paradigmatic of the naturalized biosocial making of the organism.

When evaluating the experience of genetic and chronic illness in Greece, one is primarily confronted with a highly biomedicalized reality. Conventional medical technologies are present along with gene-talk and novel biotechnologies (Papagaroufali 2002, Paxson 2004, Traka 2007, Georges 2008). Despite ambivalent feelings expressed in the media toward the latter and linked mostly to genetically modified foods and cloning (Einsiedel *et al.* 2002, Wagner *et.al.* 2002), medically related biotechnologies seem to be broadly practised and widely embraced by people and institutions. Assisted reproduction is a characteristic example. A biomedicalized therapeutic context may serve to uphold the bio/social distinction due to the infiltration of biological and genetic determinism into biomedical reasoning, placing a geneticized biological human essence at the top of a list of constituent forces in the making of the person. At the same time the dead ends together with the potentialities encountered within the context of biomedicalization are revealing, even more visibly and tangibly, the artificiality of the natural and the biological (Clarke *et al.* 2003, Rose 2007) together with the malleability of genetic material and the limitations imposed by considering the genetic level as distinct from organism and environment. We are thus witnessing a blending of sociobiological and biosocial interpretations (Rabinow 1996, 2008) and thalassaemia is paradigmatic of this process.

The often terminal physical restrictions linked to thalassaemia in combination with the biomedical advances in sustaining thalassaemic lives provide a prism through which to study the conceptual implications of the biologization, medicalization and technologization of (human) nature. It is argued that thalassaemic women and men are conceptually familiar with a theorizing of nature as built on culture and understood as practice (Rabinow 1996). Living is realized with biomedical intervention, a process that is perceived as morally right. Enabled-assisted-mediated-transfused life is a normal state of being; concerns that this process is against nature are not relevant in this context. The cooperation or rather overlap or merging of nature and culture is not contested or radical.[22] In her study of 'technologies of motherhood' in modern Greece (2004: 217–239), Paxson reflects on Strathern's work 'after nature' (1992a, 1992b), on how people have been mimicking nature in making societal forms and how nature has recently become more apparent and has ceased to be the basis for meaning in the light of assisted reproduction. According to Paxson:

> Culture and Nature have not been ideologically sequestered as separate, mimetic domains in Greece. Realizing one's nature through social practice does not require nature to be prior to, as grounds for, that practice. Rather than 'after nature', then, modern Greeks depict cultural activity as being *with* nature. .. It is notable that the Greek words for 'in practice' and 'in reality' are one and the same: *stin pragmatikótita*. What is

[22] Within biosocieties, normality and how it can be achieved in the optimal making of the person-organism is conflated with good health as informed by medical discourse and its power in making bodies and setting the boundaries of (organismic) normality (Nettleton 1995, Lock 1997, Sharp 2002, Lupton 2003, Rose 2006, 2007). The appropriation of medical knowledge, ethics and practice is evident in such societies and the optimal state of the natural is largely informed via (biomedical) normalities and pathologies.

> 'real' about human nature for Greeks is not its a priori fixity or inevitability or grounding capacity, but rather its realization through collective recognition of moral character established through behavior. (Paxson 2004: 213)[23]

This other 'realized nature' underlined by Paxson, and also realizable as I would add here, is present in the reading of thalassaemic lives and the noted normalization and embodiment of biomedical mediation. Thalassaemics are actively engaged in the making of their organisms. Many kinds of biomedical techniques are used at all levels of disease management, namely prevention, diagnosis, treatment and research. Additionally, the process of assisting thalassaemic lives has come a relatively long way and for each individual it starts from a very early age. In a sense, thalassaemics, both individually and collectively, are tuned-in and open when it comes to novel technoscience. Biomedically mediated yet normal thalassaemic lives exhibit the interplay of numerous biosocial configurations: beliefs, practices, social technologies, bioethical norms, drugs, prevention policies, surgeries, socialities, rights and duties. Biomedical achievements have been life-changing, reducing deaths and new births, mediating physiological growth and achieving fertility and reproductive capacity. Everyday life, a sense of embeddedness and a normalizing sequence of events are all intertwined with biomedical practices including regular blood transfusions, endocrinological monitoring, doctor visits, iron chelation, splenectomy and future gene therapies. All such processes, things and events mediate and are mediated by relationships, schooling and work, falling

[23] Paxson's work discusses the *téchni* (art) of urban Greek motherhood which is not presumed to be the opposite of *físi* (nature). Motherhood, she argues, 'is something to be worked at, achieved, and continuously demonstrated, contributing to gender proficiency'. Ethnographic accounts of sociality in Greece, specifically in relation to the construction of gender, have highlighted the key role of controlled behaviour in conceptualizing normative heterosexuality as well as the fact that there always seems to be room for strategic, skilful and subtle manoeuvring within the context of normal-naturalized behaviour, on condition that such doings are ethically concordant. Ethics and socially accepted and expected behaviours are well connected in this context (Papataxiarchis and Loizos 2001, Paxson 2004: 213–214). Gender construction lies at the heart of the making of the person. Understanding the making of women and men therefore catalyses the understanding of (human) nature at large. Being a woman (or a man) is not a natural thing but a 'system of virtues', something to be learned and to become good at. 'Gender proficiency' (Paxson 2004) is not a static process but a relational endeavour, consolidated, corroborated and actualized by public acceptance, by the public's gaze.

in love, economic disparities, healthcare constraints and all possible processes theorized as belonging to the social realm and involved in the contingent making of the person-organism.

Biomedically enabled lives are perceived as normal and a matter of process and not as *res judicata*. This sense of a naturalized artificiality does not imply a denial of biomedical dependence but points to the embodiment of human activities of all kinds. As often stated by thalassaemics, 'I cannot see, I cannot know what it would be like otherwise'. Awareness of bodily restrictions and awareness of dependence on other people and on technology does not threaten the integrity of the person but rather points to a bodily-cum-organismic human nature and to humanness at large as justified sites of artificiality and relationality. Human interdependence and human artificiality are brought to the surface together with the apprehension of human deficiency and lack of autonomy, of human precariousness and of the contingency of life. Thalassaemics are quite familiar with the inevitability of death; most grew up with the belief that they will die young. Having overcome fate, having experienced life-changing advances in treatment, having witnessed the death of many co-sufferers and being constantly dependent on someone else's blood for survival, they are in a unique position to acknowledge the fundamental limitations of living organisms.

The artificial-natural crafting of life in this context, besides being an embodied experience, is a result of constant management and monitoring. The early age at which a thalassaemic must confront her disease, in combination with necessarily chronic and ongoing attendance, mediates the naturalization and normalization of actions, experiences and practices otherwise possibly considered as extraneous. Among the most obvious areas of biosocial activity highlighting the normality and naturality of biomedically mediated lives in the case of thalassaemia is *the making of women and men* and the *process of blood transfusions*.

The making of women and men

> The issue of sexual maturation is very important. Most of us, we don't enter adolescence on our own. What I mean is that it will happen with the pills. To get your period, it will be with the pills. If you don't have a treatment you are going to end up with a child-like body and hence a non-sexual one. So forget about relationships! You will lose it all. But science has progressed on this issue as well. Things are looking pretty good. It is not the same when it comes to women and men. For men, if it

> shows they can't hide it easily. For women the biggest problem is fertility. But I am telling you the younger generation has no problems because they started [treatment] early on. (Arianna)

Acquiring bodily characteristics significant in constructions of womanhood and manhood constitutes in this context a process of involved action, of embodied practices, techniques and therapies, choices and availabilities. Achieving sexual maturation, having a regular menstrual cycle for women, gaining height and acquiring body and facial hair for men, being fertile, are all biomedically mediated processes and the result of skilled human action, one's own and that of others. Biomedical assistance starts early on and continues into adulthood when assisted reproduction is used, for example, in attempts to have, or rather to make, a child.

> Especially for women, if you were careful and proactive when young, if you followed the protocols and did what you had to do, then you are usually rewarded with a satisfying physical appearance, a sexy, fertile body. You have ensured your future in a way. (Costas)

Greek perceptions conflate sexuality with reproduction – for women at least – and gender normativity with the making of the person (Papataxiarchis and Loizos 2001). Thus, reproductive capacity and the making of children are of prime importance to Greek thalassaemic men and women – achieving parenthood is a ticket to normality. In the area of reproduction and not only in the case of thalassaemia, what is important is the correct use of the available technologies and not the use of technologies as against the 'natural way'. The 'making of children' is construed in terms of achievement (Paxson 2004). This conceptualizing of reproduction, of achieving motherhood and fatherhood by way of conscious action, is even more prominent and evident in the case of thalassaemia, where the process of personal involvement and use of biomedical technologies starts from early on. In other words, it is about skilled praxis and craftsmanship, about shared and exchanged, relational, socially valid and recognized knowledge.

The making of one's body in reference to gender is certainly not unidirectional nor is it solely medically mediated. Behaviours and body techniques are chosen and actualized based on the type of woman or man to be made and always considering limitations imposed by the illness. For example, the short height of women, when and if considered a problem, can be corrected with high heels, and pale skin with some make-up. Social recognition is achieved with an 'improved' appearance together with a successful career, economic prosperity, parenthood and so forth. Activation of biomedical mediation for making one's gendered self involves a weighing of options and discipline. Some thalassaemics are more involved in this process than others. Some choose to do 'whatever is needed and advised' but with varying consistency. Some women activate all possible processes and resources to achieve motherhood, overcoming fertility and conception problems and even putting their lives at risk, given that pregnancy is often risky for thalassaemic women. Others like Maria have chosen an alternative path:

> When I was 18 they invited me to the clinic and they told me I should visit the endocrinologist. They said I should go there and they would give me stuff to take. I said okay and so they scheduled an appointment. The day of the appointment came but I went swimming, I didn't care that much. And this happened again and again. I didn't really want to; I said, I don't want to have a [menstrual] cycle, I don't want to grow taller, to become sexy, that is how I felt. I want to remain little, thin and cute. If someone falls for me like this, that's fine. I didn't care that much. I have a million other things to do. In other words I didn't make an effort and I have no regrets, you know, all that, to take loads of pills, hormones, all that, for the cycle, in order to grow, for the body to grow more. Let's say to overcome a child-like body-type. But I didn't care. I decided I didn't want to take that road and I didn't.

Transfused yet normal lives

For thalassaemics, blood transfusions comprise primal and fundamental acts of biomedical mediation and embodied artificiality. Blood in this context is technologized, changed and exchanged, transferred, bought, stolen or stored. Even if in other contexts blood transfusion may question bodily boundaries, for thalassaemics it is 'a routine process', 'it is the way it is'. 'The influx of foreign blood is matter-of-fact.'

> My relation with blood is peculiar. It is an experience. I remember I liked it every time I went. Perhaps because I have a feeling of security afterwards. You feel strong. A positive experience when I think about it. I never really took the time to think about it. It can be kind of dull. You have to go to a hospital. You say to yourself: I have to do that now. Let's do it and get on with it. Sometimes it is like going shopping, you see the same things. It is normal, it is like going anyplace else. The difference after transfusion is that you feel strong. When you climb stairs, when

> you walk. Before, you feel weaker . . . It is like a flame, you feel fresher, more alive, as if you are getting up. Every time you have to do some tests, check your hematocrit, your haemoglobin levels . . . you do the transfusion and you book your next appointment. In my case it is pretty standard. I usually go every 15 days. (Panos)

> At least in my case, I don't know for the others, but I believe they feel the same, it is just like taking some kind of medicine. Like any other medicine taken this way. It makes no difference because it is blood. Except for the worrying about getting infected with something . . . if there is something in the blood . . . still it is not a constant fear. Perhaps this worry is the only thing one may be thinking about . . . You don't constantly think that someone has given their blood for you . . . It is a habit, since you start doing it so young. You are so familiar with this action. Your blood is more like a medicine you take and not something that is irreplaceable. (Katerina)

Transfusion becomes problematic and disrupts a sense of living continuity when the adequacy of the blood donation system and the correct laboratory preparation of the blood are not ensured – when, in other words, society cannot or will not care for those needing regular blood transfusions.

> It is like water for us, when there is shortage, or when it is not clean, that is when there is a problem, that is when we are in danger.

Obtaining an adequate quantity and quality of blood conditions the past, present and future of multiply transfused individuals and their families. Thalassaemics undergo blood transfusions on average every 15 days. In the past in Greece, blood had to be secured via relatives, family friends, acquaintances or even strangers – and one often had to pay for it. Nowadays, blood is primarily provided by state institutions, albeit with occasional discrepancies in quantity and quality. Thus, thalassaemics contribute to raising public awareness of the need for voluntary blood donation by mobilizing friends and family. Thalassaemic organizations actively participate in campaigns, cultural events, conferences and TV spots.

Blood transfusion symbolizes the dependent and assisted lives of thalassaemics as well as a key shift: blood becomes a symbol of craftsmanship and technicality, of voluntary donation, social interdependence, social exchange and exchange of bio-fluids and no longer symbolizes only individuality, biological relatedness and genealogy (Schneider 1969, du Boulay 1984, Koumandou 1985, Just 1989, Herzfeld 1992, 1997, Yanagisako and Delaney 1995). Adequate quantity and purity are achieved on the basis of biomedical skill but also on the basis of institutionalizing blood donation and the cultural codes that condition anonymous voluntary giving and social solidarity. Prior to any biomedical mediation, transfused blood comes from sympathetic and engaged citizens and, similarly, 'clean blood' comes from responsible and ethical donors.

Blood thus becomes a product, an outcome of human activity. In this sense, the person's authenticity and integrity are drawn away from a fixed, a priori or naturalistic point of reference. The blood running into one's veins is every time someone else's blood, it is prepared and conditioned by skilled technicians, it is cleaned and checked meticulously. It is almost totally artificial – something many thalassaemics actually wished for. Blood in this context is not a self-evident constituent of the organism. It is a therapeutic means of fundamental importance for sustaining life.

In the case of multiply transfused people, life is fundamentally linked both to biomedical knowledge and practice and to the acts of all those giving and preparing the blood for transfusion. Such an existence presents a very tangible case of relationality and biosociality, highlighting at the same time both the potential and the limitations placed on human becomings.

GENETIC (UN)CERTAINTIES

An increasingly molecular and genetically oriented medicine creates new possibilities along with new impasses. Genetic manipulation and the remaking of the (genetic) person are becoming a reality. Advances in gene therapy, especially in relation to thalassaemia,[24] have fostered hope while questioning the unchangeable nature of an inherited genetic fault. Growing knowledge of human genetic differences points to the idea that everyone, in one way or another, is at genetic risk for something, either now or in the future. The boundaries between heterogeneity, genetic difference and genetic anomaly are fluid, fuzzy and of concern to everyone – whether or not labelled as the bearer of genetic disease. At the same time, the incongruity between genetic labelling and targeted treatment leads to the formation of ambiguous

[24] Gene therapy research on thalassaemia is gaining increasing public recognition. Furthermore, in many cases Greek scientists are involved (Papanikolaou and Anagnou 2010, Yannaki, Emery and Stamatoyannopoulos 2010, Yannaki *et al.* 2012).

groupings of people often trapped between disease categories, vulnerable to deterministic or indifferent biopolicies. Gene talk, together with the revealed genetic heterogeneity, provides a pool of metaphors with which thalassaemics can attempt to make sense of their place among all other identified or imagined sources of human variation. The aim here is to discuss the severity-linked diagnostic categories of β-thalassaemia in relation to genetic heterogeneity and to consider the complexity of genotype–phenotype relationships, drawing from expert discourses and from the ways thalassaemics themselves embody context and contingency.

Biomedical knowledge constructs new dogmas and new biologies with direct or indirect implications for individuals, groups and populations. The shift from gene-centrism towards more context-sensitive approaches, which is taking place at a different pace in basic research, clinical research and clinical practice, has a variety of meanings in people's lives and is filtered by local notions and institutions. Apart from impacting on diagnosis, the discovery of new mutations produces discourses of certainty that are translated into prevention practices, as in the case of thalassaemia. At the same time, the growing awareness of the complexity inherent in attempts to correlate a genetic template with a living organism – as in genotype–phenotype thinking – leads to uncertainty. New disease or at-risk groups are created.

In the case of β-thalassaemia, the established categories of major, intermedia and minor, linked primarily to levels of anaemia and the need for transfusions, have been used to define specific treatment protocols.[25] The basic indicator of clinical severity is the reduced or absent synthesis of beta globin chains and the extent of the imbalance between alpha and non-alpha globin chains (including gamma chains, variably present in β-thalassaemia syndromes). Patients with β-thalassaemia minor are usually heterozygotes and clinically asymptomatic with the exception of certain dominant β-thalassaemias linked to rare mutations. Homozygotes develop into either major or intermedia, with the former requiring transfusions as early as at 2 years old to survive and the latter having a very heterogeneous clinical picture. It is often difficult to differentiate between major and intermedia. In such cases, a genotype analysis is advised at the alpha and beta loci as well as testing

[25] Some other less prevalent forms are due to the interaction of HbE and β-thalassaemias resulting in phenotypes similar to major and mild intermedia and some rare cases where the defect does not lie in the beta globin gene cluster but is rather associated with other features (Galanello and Origa 2010).

for the presence of ameliorating genetic factors (Cao and Galanello 2010). Intermedia ranges from mild cases which are asymptomatic until adult life, with only mild anaemia, to very severe cases which present between the ages of 2 and 6, in patients variably capable of surviving without regular transfusions, growth and development complications (Galanello and Origa 2010). A brief account of the logic of the genetic aetiology highlights the documented genetic heterogeneity, the difficulty of correlating genotype and phenotype and the consequent difficulty in understanding the mechanisms involved, even in this monogenic disease.

In trying to understand clinical–molecular links, the key pathophysiological determinant informing severity is the extent of alpha/non-alpha globin chain imbalance and any factors capable of affecting this relationship (Weatherall 2001, Cao and Galanello 2010). A plethora of such factors have been identified, which may accentuate or ameliorate severity by way of straightforward or complex molecular mechanisms. The list includes a variety of primary mutations of the beta-globin cluster, secondary modifying loci and tertiary modifiers as well as the effects of co-evolution, environment and culture (Weatherall 2001). Thus, in addition to the more than two hundred recognized mutations of the beta globin gene that relate to various molecular outcomes, other genetic variants seem to be involved in the disease phenotypes, triggering different molecular mechanisms which variously modify severity (such as co-inheritance of α-thalassaemia, deletions of the 5′ region of the beta globin gene and polymorphism of the BCL11A gene). Co-inheritance of genetic variation outside the beta globin cluster including, for example, the apolipoprotein E ε4 allele, some HLA haplotypes, genes involved in iron metabolism and genes involved in bone metabolism, may further complicate the phenotypic outcome (Cao and Galanello 2010, Galanello and Origa 2010).

Information regarding heterogeneity is increasingly available due to the positive effects of better disease management that significantly increases thalassaemic life expectancy rates. Research on a growing number of adult thalassaemics has revealed additional sites and mechanisms of complexity and disease progression. This eventuality further points to the processual, relational and biosocial nature of the development of the person-organism. Biology is revealed here as a developmental process and not as mere genetics (Ingold 2001a), linked to the environmental conditions of the organism at every moment in time. Even in the case of usually asymptomatic heterozygotes, several environmental or genetic factors are believed to alter the phenotype.

Phenotypes may fluctuate throughout one's life via factors ranging from infections and nutrition to more specific problems such as splenomegaly linked to the unfolding of the condition itself. The relationship between high frequency of β-thalassaemia in certain parts of the world and a heterozygotic advantage against *P. falciparum* malaria further elucidates the expansion of polymorphisms at genetic loci other than the beta globin cluster including HLA, TNF, ICMA1, tumour necrosis factor alpha and intracellular adhesion molecule-1. Such polymorphisms play an important role in defence mechanisms against infectious diseases, explaining cases of geographically variable biosocial responses to infection (Hill, Allsopp and Kwiatkowsi 1991, McGuire *et al.* 1994, Fernandez-Reyes *et al.* 1997, Weatherall, Clegg and Kwiatkowsi 1997, Cao and Galanello 2010: 68). The formation of adult thalassaemic populations in different parts of the world makes possible comparative research revealing the role of the environment (Weatherall 1998, Weatherall and Clegg 2001) while calling for more holistic and inclusive approaches to understanding how the disease unfolds. Cases of the same category (minor, major or intermedia) may be linked to different genotypes. This complexity and heterogeneity make it increasingly difficult to think in terms of clearly defined phenotypic subcategories, to the extent that the general distinction between monogenic (supposedly more straightforward) and multifactorial (supposedly more complex) diseases is becoming problematic. Such distinctions have a history and are bound to be questioned on the basis of more recent knowledge.

Thus, β-thalassaemia furnishes an exemplary instance with which to evaluate gene-centric approaches and the essentialization of biology. The more we know about the genetic level, the genotype-phenotype relationship, the co-inheritance of various genetic variants and the clinical picture, the more apparent the overlap of all these levels of investigation becomes. When we speak of clinical phenotypes, we should really mean the person-organism living in an environment. The logic of the disease aetiology which starts from the most well-documented mutations linked to the beta globin gene cluster, followed by the differentially situated and variously interacting genetic modifiers of the alpha and non-alpha chain imbalance (primary, secondary, tertiary), further followed by acquired and environmental factors, gives undue weight to the genes and supports the idea that they underwrite 'layer upon layer of complexity' (Weatherall 2001). This bias, even if unintentional, is the product of a way of thinking according to which multiple (culturally) acquired mechanisms modify the expression of (naturally) given genes. The example of β-thalassaemia is exemplary in revealing the inadequacy of such conceptual tools, of the duality of genotype and phenotype, and of categories such as 'major', 'intermedia' and 'minor'. It is no accident that in the case of thalassaemia, an increasing appreciation of heterogeneity and complexity has been followed by a turn towards individualized assessment and treatment. Weatherall has stressed on many occasions the need to look at the bigger picture:

> The principal lesson from the thalassaemia field is that, before it is possible to start making real sense of phenotypic diversity, it is very important to establish the precise pathophysiological mechanism of the disease. So, although it might be possible to identify modifiers by sophisticated genome searches, without an understanding of the cellular pathology of a disease it will be very difficult to determine the significance of putative loci that are ascertained in this way. These observations provide some indication of the difficulties that will be encountered when trying to identify the major genes involved in susceptibility to common multifactorial disorders, such as heart disease, hypertension, diabetes, the common psychoses and so on. If this still primitive understanding of the mechanisms for the phenotypic diversity of the β-thalassaemias has told us anything, it is that it will be absolutely vital accurately to define phenotypes and broad pathological mechanisms at the same time as embarking on a search of the genome for susceptibility loci. If this is not done, many candidate genes that are no more than very minor phenotypic modifiers might be mistaken for important players in the basic pathogenesis of these complex diseases. (Weatherall 2001: 253)

Returning to the ethnographic field, diagnosis and prevention of thalassaemia are the two areas most vulnerable to bio-determinisms (phenotypic and genetic) and to the uncertainty produced by increasing awareness of underlying complexity. Given that thalassaemia is one of the best-studied of monogenic diseases, which means that both experts and non-experts have long known its inherited nature, the more recent emphasis on genetic aetiology, along with the potential geneticisation of disease management, has not been as sudden as with other diseases. Managing thalassaemia has been based to a great extent on clinical manifestations and has led to the establishment of the disease subcategories mentioned above. These categories inform treatment guidelines that are important for the developmental unfolding of thalassaemic lives, interacting in manifold ways with the making of the clinical phenotype. In the case of thalassaemia major, the fixing of categories like major and intermedia, as well as the established category of 'multitransfused' patients (which may include major and intermedia cases), has serious implications when it comes to accessing state benefits. In general, fixing categories based on clinical phenotypes and more recently on genetic markers may lead to exclusions, not only from state-supported healthcare benefits but also from active patient groups. New knowledge is often translated slowly into clinical practice whereas labels and categories based on either genotypes or phenotypes tend to persist despite their often ambiguous status, contributing to the formation of the biosocial person.

Following the more recent routinization of genetic testing in diagnostic practice, one observes a sense of ambivalence especially on the part of patients themselves. Scientific and clinical attention to the molecular level often means a shift from symptomatology to genetic causation, and from the need for blood to genetic faults and mutations. This growing interest in the genetic level, together with the pre-existing subcategorizing of β-thalassaemia, seems to be having two broad effects. First, information regarding the molecular level is used together with clinical data to draw conclusions about severity – via genotype-phenotype correlations – leading to further subcategorization, such as the subdivision of intermedia into 'mild', 'moderate' and 'severe' (Taher *et al.* 2009), and to changes in treatment. The emergence of new diagnostic labels and new disease subtypes, despite their messiness, creates tensions and re-evaluations of who belongs with whom (of the same or different genotype), affecting the formation of biosocial subjectivities (Chatjouli 2012). Secondly, the molecularization of diagnosis and disease mapping is infiltrating policies regarding access to benefits and care, further constructing genetic citizenship. Thus, despite the growing awareness of the complexity involved when attempting to relate findings at the molecular level to clinical phenotypes, the overall geneticization of medical practice has significant implications not only for the handling of patients and therefore for the development of the person-organism via treatment, access to benefits and inclusion in patient groups, but also across generations through prevention programmes, population screening and the like.

The difficulty faced by experts in their attempts to organize the remarkable phenotypic diversity of β-thalassaemias comes as no surprise to thalassaemics who experience ambiguity on a daily basis. Petros, an adult male diagnosed with thalassaemia intermedia, uses the metaphor of the curve to talk about stages of severity, arguing that the clustering of in-between or confusing phenotypes into the intermedia category can be problematic when reflected in lived experience. Belonging to the intermedia category often means too many different things, thereby making categorization meaningless. The diagnosis results in specific therapeutic guidelines, further affecting the process of severity-in-the-making. Petros stressed that while severity is mostly measured by the need for blood, there are numerous intermedia-typed thalassaemics who have regular transfusions. Moreover, plenty of intermedia cases known to Petros led more difficult lives with poorer health primarily due to a regime of less aggressive disease management which, for example, focused on palliation rather than on preventing anaemia-induced complications.[26] Petros also recalled sufferers labelled 'severe' who survived without regular transfusions for an unusually long period, as well as other 'severe' cases whose overall health was better compared with intermedia cases, probably due to timely monitoring and treatment. Perhaps the most poignant example is that of Petros's two sisters, who were diagnosed as less severe cases than himself but are no longer living.

> There is so much diversity with thalassaemia, even in the case of being a heterozygote; it is not the same . . . In other words there are 250 genotypes and each one has a different behaviour to the illness. Being as they say a heterozygote or a homozygote is not something specific for all. Some have it to a high and some to a lower degree. Think about it as a curve . . . it is not about having it or not having it . . . there are gradations and different genes interact with each other and give different outcomes, like in the intermedia case. They tell you that you have thalassaemia but you don't have transfusions, only when needed. Me, they say, you have intermedia, but I have regular transfusions. Also my sister, she got transfusions when she was nine.
>
> Well, it is the behaviour of thalassaemia [that is] different for each person. You cannot foresee for each person how many years she would live, or you cannot [predict] the body [and its reactions]. That is why you see so many different situations . . . It is like that, like throwing the mother's and the father's genes into a bucket, you don't know what will come out, but the new will never be the same. (Petros)

Using rigid categories of severity based on symptomatology or genetic typing for diagnosis seems to reduce the potentiality of the person-organism as a whole. Acknowledgement that the environment should

[26] There is a growing literature questioning the established management of the intermedia category, often arguing in favour of individualized treatments (Aessopos *et al.* 2007, Cappellini *et al.* 2009, Taher *et al.* 2011).

be included in understanding the documented variation has not yet been adequately translated into disease management. Being labelled with a specific diagnostic category usually from an early age (if not prenatally nowadays) has direct and indirect implications for the unfolding of the condition. According to Petros, one's position on the 'severity curve' fluctuates throughout life.

Different diagnostic categories inform the construction of the boundary between biological normality and abnormality, and between biological abnormality and biological diversity, with implications for the biopolitics of prevention. Many thalassaemics, aware that these boundaries are shifting, have raised their concerns regarding how preventative decisions are made. Which cases are severe enough to become prevention targets, which phenotypes and which genotypes? How often should such decisions be revised? Is all this ambiguity represented in genetic counselling sessions?

> As I was telling you, we know from historical findings that thalassaemia has a very long history . . . Now think of the intermedia case. I know a girl who started transfusions when she was around 25. . . . There is another one in Crete, he is 70 years old. I told you if you have two mild genes you [have] a very mild form of thalassaemia. Many people lived with low levels of haemoglobin, like our parents. So, there are cases [in which experts] would say you must abort the child but which are quite viable. Now, I don't know to what extent science can be very precise and make the exact prognosis before birth looking at the genotype alone. I don't know to what extent this can be done.

The increasing emphasis on the genotype together with science's ability to see it and even to change it has led to an understanding and management of the genotype more and more similar to that of the phenotype – even to the conflation of the two (Palsson 2007a: 230). In a way, the genotype has been exposed, has come to the surface, it has been shifted from its role as a bearer of coded information and been essentialized. The problem is that both genotype and phenotype are decontextualized. Both concepts are problematic, not only in relation to each other but also as biological centres of information or, even more problematically, as biology themselves. According to Ingold's critique of the concept of the genotype, 'there is no reading of the genetic code that is not itself part of the process of development' (2001a: 260–261). Revealing the problematic assumptions underwriting the genotype/phenotype distinction, whereby individuals are specified in their essential genetic constitution – independently and in advance of their life in the world, through the bestowal of attributes from ancestors – Ingold highlights the need to resituate persons-organisms in their environments.

BIOLOGICAL DIFFERENCE IS RELATIONAL AND BIOSOCIAL

The thalassaemic paradigm challenges bio-genetic and social reductions of lived experience. It forces us to question not only various forms of genetic determinism but also various forms of a fragmented biology, as in the genotype-phenotype model. Such reductions and fragmentations often fail to do justice to the polyphony of human life, to the living flow of relational accomplishment and the involvement of the self in the process of becoming (Ingold 2010c). The emphasis placed by thalassaemics on 'being at home with their bodies', which are under constant mediation (biomedical and other), reflects the embodiment of human activities and everyday experience (Ingold 2001a) as well as the irreducibility of biology to genes, pointing to a conflation of the biological and the social grounded in a lived sense of the relational, non-autonomous, continuous, complex and contingent making of life. Perhaps this very talk of the organismic self reflects how these discourses, applied to the body, highlight its instability (Palsson 2007a: 229–230), as well as the instability of biology and nature and the need to relate them to local contexts and conceptions.

Biological difference, identified on both genetic and physical levels, is a hard fact with which thalassaemic individuals have to come to terms. Their bodies – pathologized, geneticized and technologized – become sites of constant biomedical treatment and experimentation. As recipients of the changes brought by hypes, trends and shifts of biomedical knowledge, they experience the immutability together with the plasticity of biological natures as well as the constructed natures of biological truisms. Within such an environment, thalassaemics make sense of the world by 'being thalassaemics and having thalassaemia', unable to separate the two, 'unable to imagine another state of being' as they often state. With their natures constantly at stake, with their highly technologized bodies, with their evident dependence on human activity and on other people's blood, thalassaemics reveal in the crafting of their lives how biology itself is irreducible to genetics and intrinsically tied to human action, as a developmental and relational process, as constantly in the making.

NOA VAISMAN

6

Shedding our selves: perspectivism, the bounded subject and the nature–culture divide

It is a well-rehearsed claim that the modern western conception of the human is premised on the ontological distinction between nature and culture (Franklin 2003, Haraway 2008). Nature is the undifferentiated foundation upon which culture is applied, and culture is what distinguishes and differentiates us from other animals. This opposition, as Palsson points out (2006: 73), emerged with the advent of modernism, and rapidly became a key component in our conceptualization of the human. While in medieval Europe the person was viewed as entirely embedded in a nature-culture world, in the western modernist ontology the person is envisioned as the product of a culturally specific process of socialization. Thus, as Ingold describes this view, the person is 'made through the imposition of a specific cultural form upon a pre-existing and undifferentiated material substrate' (2006a: 181). While ethnographic examples of non-western peoples have provided a clear indication that the nature–culture divide and other related binaries are not universal, the critical examination of the distinction between the social and the biological within western ontology has only recently gained force (Latour 1993, Descola and Palsson 1996, McLean 2009).

In part, this critique is the result of our growing ability to intervene and change nature according to our needs and desires by manipulating genetic material and producing machine-human interfaces. For example, new reproductive technologies (NRTs) such as ova and sperm donation, surrogate motherhood, and IVF have challenged us to reconsider the 'naturalness' of biological processes such as insemination and

Biosocial Becomings: Integrating Social and Biological Anthropology, eds T. Ingold and G. Palsson. Published by Cambridge University Press.

parent-child relations (Strathern 2005); meanwhile, tissue cultivation and organ transplanting have blurred the boundaries between humans and between humans and other species. In most general terms, these technologies have redrawn the map of 'life', and in the process called into question such basic dichotomies as nature/culture, subject/object and organism/person. They have critically challenged many of our taken-for-granted assumptions about human nature and the relations between humans and their environment. And they have also given rise to new ways of thinking about the nature/culture divide. For example, Haraway's work and the concepts she has coined such as 'cyborg' and 'natureculture', and initiatives such as the multispecies salons developed by anthropologists Kirksey and Helmreich (2010), have urged us to acknowledge the intermingling and intertwining of humans, animals, microbes and machines. But while these works have attempted to dissolve the nature/culture dichotomy, they have in most cases ended up endorsing one of two positions, both of which are still premised on the distinction between nature and culture: either nature is shaping culture or culture imposes meaning on nature.

In the following discussion, I draw on what I consider to be a more critical view of the nature–culture divide, specifically Viveiros de Castro's reading of perspectivism in Amerindian thought. Reflecting on a wide range of ethnographic materials predominantly from Amazonian societies, Viveiros de Castro explains that what has been identified as the 'perspectival qualities' of these societies lie in their understanding that the world is inhabited by 'different sorts of subjects or persons, human and non-human, that apprehend reality from distinct points of view' (2012: 45). 'Humans see humans as humans and animals as animals', but animals see themselves as humans, they see their food as human food, they see their social system as organized in the same way as human culture, and they see humans as animals or prey. 'For Amazonian peoples', he writes, 'the original common condition of both humans and animals is not animality but, rather, humanity' (2004: 465). What unites different animal species (humans among them) is culture, while what differentiates between them is their nature. Perspectivism, Viveiros de Castro goes on to argue, cannot be explained as just another interpretation of reality – the native's worldview – but is rather an expression of different worlds, or in his terms, different ontologies. Amerindian ontologies are based on a spiritual unity in that we (humans and some animals) are all persons, and on a natural diversity in that we all possess different bodies, a point he explains further by positing the concept of 'multinaturalism' against the modern western concept of 'multiculturalism'. Different bodies, then, produce different perspectives. Nature and culture in Amerindian thought are not therefore ontological domains but rather products of relational and exchangeable positions. They are the outcomes of different points of view. It is not that the nature–culture divide is wrong or obsolete, but rather that Amerindian ontologies offer us new and different perspectives on our much entrenched dichotomies (2012: 47).

Drawing on Viveiros de Castro's reading of ethnographic accounts of the Amazonian region and on his notion of ontologies and multinatures, in this chapter I consider a 2009 Argentine Supreme Court decision in a case involving identity tests on shed-DNA. The central question the Court deliberated was whether a DNA identity test using shed-DNA – that is, DNA that can be collected and extracted from biological materials left on objects in the environment inhabited by a person – should be carried out against an individual's expressed wishes for the purpose of discovering his 'true' identity. The chapter is a companion piece to a more socio-legally oriented analysis of the same case. While in the present chapter I explore the Supreme Court's view of the subject's embodied existence, the companion piece questions the use of new scientific technologies, such as shed-DNA identity tests, in the context of legal deliberations of ongoing crimes against human rights.

THE CASE

During the last military dictatorship in Argentina (1976–1983), some 500 infants and small children were abducted by the regime. Once separated from their birth families, they were provided with new identity papers that indicated kinship ties with their appropriators, and were given new names and dates of birth. Through these means and the silent approval of the social worlds they inhabited, these individuals were never made aware of their actual birth origins or of the search for them conducted by their birth families (see Abuelas de Plaza de Mayo 2008). Today, these individuals are in their early 30s and are regarded as the 'living disappeared' of Argentina (Arditti, 1999). Guillermo Gabriel Prieto, whose case in front of the Supreme Court of Argentina I analyse in this chapter, was presumed to be one of these individuals. He was thought to be the child of María Ester Peralta and Oscar Alfredo Zalazar, both forcibly disappeared in the early years of the military rule. His presumed birth mother was five months pregnant when she was abducted from Villa 21 de Barracas. She gave birth in one of the clandestine camps where those disappeared were detained, tortured and often eventually assassinated. Shortly after his birth, the infant was given to Guillermo Antonio Prieto, then a non-commissioned officer in the Navy, and Emma Elidia Gualtieri, his wife. He was raised as their 'natural' son along with his brother Emiliano Prieto.[27]

Over the years suspicions regarding Guillermo's identity mounted. In a legal investigation in 1992, the Prieto–Gualtieri couple declared that Emiliano and Guillermo were their biological children, both born in a private clinic, but the courts were not satisfied with these declarations, and requested that Guillermo provide a blood sample. The blood was to be used to clarify whether Guillermo was in fact the couple's child or the child of the disappeared Peralta–Salazar couple. The appropriating couple 'systematically resisted'[28] these attempts, and when Guillermo reached adulthood, he himself refused the blood test. Guillermo's resistance led him as far as the Supreme Court of Argentina, the federal tribunal of last resort. While this case was being debated, new technologies became available and a judge in a lower level court ordered and presided over a search in Guillermo's house. During this raid, Guillermo's personal objects were collected with the aim of gathering enough biological material to carry out a DNA identity test. The Supreme Court decision centres on the extraordinary appeal that Guillermo filed against this search and the testing of his shed-DNA.

The Court's ruling begins by sketching out Guillermo's claims in his extraordinary appeal as well as in earlier instances; specifically, that the collection of shed-DNA in the house raid is no different from the blood test since both have the same goal, which is to carry out an identity test using his DNA. Second, the use of his biological material against his wishes violates a number of his constitutional rights, forcing him to question his own identity and compelling him to set himself

[27] The children's birth certificates indicate Guillermo Prieto and Emma Gualtieri as their natural parents. However, it was assumed that Emiliano and Guillermo are not genetically related; rather each was abducted from a different couple during the military dictatorship. Emiliano's presumably falsified birth certificate states that he was born in August 1978 while Guillermo's states that he was born in June 1976 (*Página 12*, 11 November 2009).

[28] All quotes are from the Supreme Court of Argentina case G. 291. XLIII Gualtieri Rugnone de Prieto, Emma Elidia y otros s/sustracción de menores de 10 años. All translations of the Supreme Court documents are my own.

up as evidence (in the penal trial) against his parents.[29] The ruling then offers a detailed response to these claims and repositions the case within a broader local and global socio-legal context. While acknowledging Guillermo's position as a victim of abduction and alteration of identity, the Court identifies his presumed biological family and particularly his grandmother, Petrona Catalina Izaguirre de Peralta (María Ester Peralta's mother), as victims as well. It argues that the family has the right to find out if Guillermo is their kin and to advance in the investigation of María Peralta's forced disappearance. Guillermo, on the other hand, has a right to his intimacy, his privacy, his physical integrity, as well as a right to property in his genetic material. From this depiction of the case, it becomes apparent that these competing rights point towards very different solutions. The Court's task, the Justices state, is to find a 'point of equilibrium' between the two positions.

The Court resolves the case in the following manner: First, it states that the protection of rights delineated in the Constitution is not unconditional. In effect, the right to privacy and intimacy can be violated when the state is required to clarify the truth in a penal case (G. 291. XLIII: 28). Moreover, since this case is intimately tied to an instance of 'forced disappearance' (the disappearance of María Peralta and Alfredo Zalazar), which has been recognized by the courts as a crime against humanity, the state has both a local and an international obligation to investigate and find the truth about the crime. Second, the Court states that:

> It is not observed that the method in question [shed-DNA identity test] would cause damage to fundamental rights such as the right to life, to health, to corporeal integrity or intimacy, since the specimens were taken without invading the body of the appellant, furthermore, even without his active participation, and its utilization is aimed at guarding public interest, which is demanding the determination of the truth in the case. (G. 291. XLIII: 29)

The biological material, the Court goes on to note, 'while belonging to his [the appellant's] body, had been detached from it at the moment of confiscation' (G. 291. XLIII 2012: 30). This distinction between the body and its traces allows the Court to further claim that 'anything that

[29] 'The resolution . . . ordered against his wishes the use of his biological material which is his property, forcing him in this way to constitute himself as evidence against his parents and affecting his dignity, therefore forcing him to question his identity' (G. 291. XLIII: 22).

emerges out of the biological ties that may be determined through the test is reduced exclusively to the private and affective terrain. In this way, its directing and its external manifestation are unconnected to any decision or interference of the courts' (G. 291. XLIII: 29). In other words, the Court distinguishes not only between the body and its parts (that is, shed-DNA and the whole body) but also between the elucidation of the 'truth' in the case and the emotional and relational outcomes of the knowledge that will emerge from the test, as well as the ties that may possibly develop subsequently.[30] Finally, in line with the above claims, the Court states that the blood test cannot be forced on the victim, but that the shed-DNA identity test can be carried out on the materials collected during the house raid. This DNA identity test would be the best and only adequate method of arriving at the 'material truth, taking into account the high level of certainty it affords' (G. 291. XLIII: 30).

The Court's decision is curious on a number of counts. First, the distinction it draws between DNA that can be extracted from a blood sample and shed-DNA raises questions about the subject's physical boundaries. Specifically, DNA inside the body belongs to the individual while DNA outside the body (shed-DNA) can be appropriated by the state.[31] Second, the Court's view is that the knowledge gained through the test can be separated from its outcome – the impact the test might have on Guillermo and his relational world. Third, the 'material truth' in the case is restricted to the appellant's DNA and the information it holds about his relationships. That is, the Court offers a rather limited view of what truth may be and how this truth may reflect upon the individual subject. In what follows, I consider these aspects of the case and work my argument in two directions: first, an examination of the embodied nature of the subject as it appears in the ruling; second, an inquiry into the kind of subject and the kind of ontology the Court's ruling produces.

[30] Here it may be particularly important to note, along with Strathern (1999), that information gives a person an identity: 'knowing something about one's kin is also knowing something about oneself – if one has no option but to deal with the information it is also true that one has no option over the relationships' (1999: 77).

[31] Joh (2011) has argued against this interpretation of 'abandoned' DNA and has called for a rewriting of legal codes to reflect the sensitive nature of shed-DNA as an indicator of our selves and our relations with others (such as familial genetic ties that can be discovered through biobanks).

UNDOING THE BODY'S BOUNDARIES

In the ruling, the Court makes a distinction between Guillermo's shed-DNA and the DNA that can be extracted from a blood sample. It observes that 'the DNA test was taken based on elements that, although belonging to his [Guillermo's] body, at the moment of collection had been detached from it' (G. 291. XLIII: 30).[32] It further states that 'the samples were taken without invading the body of the appellant' (G. 291. XLIII: 29); in fact, he was not present during the house search. In these statements, the Court is enunciating a distinction between the whole body and its parts. The whole body is understood to be inviolable, and as such un-own-able, making forced blood tests untenable. However, parts of the body (tissues that were detached from the body and left on objects in the subject's surroundings) are considered things – that is, objects that are own-able and open to confiscation. This distinction reifies parts of the body while maintaining that the body as a whole is the individual's personal realm, which the state is restricted from entering.

The particularly confounding aspect of this distinction is that (as Guillermo himself claims) the goal of both tests is to identify the individual. And while the shed-DNA is constructed as belonging to the body by virtue of its ability to provide information about it (the same information that can be 'extracted' from the body itself), it is also not part of the body because it had detached itself from it. Herein lies the confusion: the body depicted by the Court requires that we look twice, once at the body as a whole – the material body of the individual as well as his shed-DNA – and a second time at the body as made up of two slightly overlapping circles where shed-DNA can be counted as both part of the body and as a separate entity that has a distinct legal value. In elaborating on this distinction I want to highlight a point not made by the Court or Guillermo in his appeal: that is, what becomes undone in the process of distinguishing between the body and its parts

[32] A very similar argument is also used in Emiliano's case where the Court acknowledges the existence of a personal space within which the individual has the right to make his own decisions and live according to the lifestyle that he chooses. Referring to the obligatory blood tests as spectacles that should be avoided, the Court states that the forced tests would violate an individual's right to privacy as it is recognized in the Argentine Constitution. Here the body appears to belong to the person – he has rights over his bounded physical being, and any violation of its boundaries inevitably causes a violation of basic rights.

is not so much the right to ownership over one's body and its parts, but rather the physical boundaries of the person himself.[33]

In effect, the ability of the lower court to order and preside over a house raid where DNA was collected implies that the clear physical boundaries that we hold to exist, in our modern western ontology, are much more porous and blurred. The body does not end at the skin; rather it extends, becoming thoroughly entangled with its environment. This way of viewing the subject and its embodied nature reminds us of what Ingold calls 'relational thinking', which means: 'treating the organism not as a discrete pre-specified entity but as a particular locus of growth and development within a continuous field of relationships. It is a field that *unfolds* in the life activities of organisms and that is *enfolded* in their specific morphologies' (Ingold 2004: 219). Ingold critiques current views of evolution and of the development of the organism as a discrete entity whose DNA encodes its growth path, and argues instead for a more comprehensive view of the human-organism founded in a relational rather than a genealogical model (Ingold 2000: 132–151). My point here has much in common with his view, though I believe it extends his notion of becoming further. Thus, the organism-human is not only a becoming, the embodiment of processes of *enfolding* and *unfolding relations* between a particular human and his or her specific historical and environmental circumstances. More than that, I argue that our originary point of reference – the organism as a physically bounded entity – may be flawed. In other words, it may be that we need to think of the human not only as organism-human but as an assemblage of environment-organism-human: a human whose boundaries are always open to the world and whose existence is enmeshed with its surroundings.[34]

[33] In this way, the case is different from other cases focusing on the question of ownership of body parts because the crux of the matter is not the actual ownership over the material or the scientific knowledge derived from it, but rather the information that can be found in these genetic materials about the person himself (see Gold 1995, Pottage 2001). This may explain why the Justices refuse to accept Guillermo's use of the Universal Declaration on Human Genome and Human Rights and his claim that the genetic material found on personal objects belongs to him and must be protected from state intervention (see particularly articles 9, 10, 25 of the Declaration). The assumption underlying his claim is that biological material, whether inside or outside the material body, is the property of the individual, and only he has rights over it.

[34] While there are many points of convergence between this view of the human and Ingold's depiction of the indigenous person who emerges through reciprocal relations with land, my point here is to stress the porous nature of the person's

While provocative, this position raises many complex questions about the subject and his or her relations to other subjects, as well as about one's rights and the nature of one's entanglements in the environment. Nonetheless, I believe it is a worthwhile thought exercise, not least because it somewhat inadvertently shakes what has become a foundational aspect of our current thinking about personhood – specifically, that we are our DNA.[35] For example, it may be significant to note that nowhere in the Court's ruling or in the Justices' representation of Guillermo's position is the direct tie between DNA and identity questioned. Instead, it seems that even Guillermo himself is resigned to equating the two, asserting that the DNA identity test is forcing him to question his identity. My claim, then, is that 'shed' DNA is in fact not shed at all but rather an extension of our body-self, which exists in and through our environment. This way of thinking about shed-DNA requires that we begin to reconsider such basic western dichotomies as inside/outside, human/environment, us/other.

A DIFFERENT PERSPECTIVE

In the preceding discussion, I suggested that shed-DNA could point us (and the Justices if they so chose) towards a vision of the subject as entangled and embedded in his or her surroundings, a subject whose physical non-boundedness gestures toward a world where such dichotomies as inside/outside, organism/environment, us/other do not exist. In what follows, I take a similar analytic path but approach the question from a slightly different angle, focusing on the Court's view of the subject. Specifically I ask: what kind of a subject does the Court produce in its ruling? In working through this question I am guided by a claim made by Alain Pottage in his introduction to the volume *Law, Anthropology and the Constitution of the Social* (2004). Pottage outlines how the fabrication of persons and things in law has been changing. Until recently, within the western legal system, the distinction was clear: things (*res*) were objects open to commodification and

physical existence per se. That is, in my reading, Ingold's relational model is still somewhat premised on the notion that the person is a bounded entity, whereas what I am suggesting here is that these boundaries need to be reconsidered, even among modern western peoples, as permeable.

35 Here it may be sufficient to point to the innumerable ancestry-tracing companies that use DNA to tie individuals to their purported origins (e.g. Elliot and Brodwin 2002, Palsson 2007a: 59–90), as well as recent discussions regarding race, ethnicity and genetics (Palsson 2007b).

ownership whereas persons were sacred, located outside the realm of transactions and property rights (excluding slavery of course). While the distinction, like that between nature and culture, is foundational to western thinking, in recent years and partly in response to technological and scientific advances, it has become blurred. Now, Pottage suggests, genetic material and biological tissue can become, in the law, both persons and things (2004: 4–5). This insight leads him to claim that 'techniques of personification and reification are constitutive rather than declaratory of the ontology upon which they are based' (2004: 9). In other words, the definition of something or someone as a person or a thing is not a reflection of a particular ontology, but rather a way of *producing* an ontology within which the thing or person can exist. This reversal is important for my argument since I take it to mean that in the Court's constitution of the subject, it is also producing (rather than declaring or reflecting) an ontology. And as I shall show, I believe that the Court is creating not one but two ontologies.

To situate this argument I want to return to Viveiros de Castro's analysis of Amazonian ethnographies and to his claim that according to Amerindian thought, 'All beings perceive (represent) the world in the same way. What varies is the world which they see' (2004: 471–472). The world, according to Amerindian cosmologies, is determined by the perspective, and different perspectives are the result of different bodies; thus 'animals *see* the same way we do *different* things because their bodies are different from ours' (1998: 478).[36] By bodies, Viveiros de Castro is referring here to affects, dispositions and capacities that constitute a *habitus* rather than to the visible shape of the body. The reason for this particular use of the term is that in Amerindian thought, although the visible body can be a good indication of perspective, it can also be deceptive as bodies (like envelopes or masks) can be taken off or put on. In sum, the point is that different bodies create different perspectives, and a change in perspective produces a different world. Or, as Strathern (following Viveiros de Castro) states, a perspective 'is

[36] Viveiros de Castro provides many examples for this claim throughout his writings and lectures: 'animals and spirits see themselves as humans: they perceive themselves as (or become) anthropomorphic beings when they are in their own houses . . . and they experience their own habits or characteristics in the form of culture . . . Jaguars see blood as manioc beer . . . they see their bodily attributes (fur, feathers, claws, beaks) as body decorations or cultural instruments' (2012: 47–48).

an ontological way of being that alters the conditions of the person ("creates the subject") so that anything in the perceived world (tangible or intangible) defines and contributes to that state of being' (Strathern 2005: 153). Picking up on these insights, I will argue next that in its treatment of the subject, and more specifically the subject's body, the Court's ruling produces two different perspectives that are two different ontologies.

The Supreme Court's ruling seems to oscillate between two visions of the subject: the subject as a product of genetic ties and the subject as a product of his or her social world. The view of the subject as a product of genetic ancestry is revealed in the terms the Court uses to describe Guillermo's appeal against the shed-DNA test. For example, 'biological identity' [*identidad biológica*] (G. 291. XLIII: 47), 'his alleged parents' [*sus supuestos padres*] (G. 291. XLIII: 33) and 'material truth' [*verdad material*] (G. 291. XLIII: 30) are terms used, respectively, when referring to Guillermo's appropriators, his identity or the person that may emerge from the test. The equation of the subject with his DNA is also prevalent in the manner in which the Court treats Guillermo's claims against the test. For example, the Court explains that the test forces Guillermo to question his identity (G. 291. XLIII: 24), and the investigation more generally presents (in Guillermo's experience) a 'constant threat against his personhood' (G. 291. XLIII: 47). In this depiction, the shed-DNA is viewed as part of the subject and a reflection of his selfhood. Further indications of this position can be gained by examining the Court's statements about the biological family. For example, in their partial dissent, Justices Lorenzetti and Zafarroni explain that apart from the forced disappearance of their closest kin, the biological family must endure the speculation or even the knowledge that there is a grandchild, a brother or a niece who is living somewhere out there without knowing about his genetic kinship ties (G. 291. XLIII: 68). For these Justices, upholding the biological family's right to the truth implies conducting a DNA test because 'this practice will permit arriving at the objective truth of the events ... that is, to determine if Guillermo Gabriel Prieto is in effect the son of the Prieto–Gualtieri couple or if he has biological relations to the family Peralta–Zalazar' (G. 291. XLIII: 30). In this way, from these Justices' perspective, DNA is tantamount to the truth, and this truth is the individual's identity.

While recognizing the intimate relations between DNA and identity, the Justices also depict the subject as a product of his relational world – that is, of the social ties that he had grown up with and developed within.[37] For example, Justice Petracchi states that in other cases when the legislator refused to accept the testimony of people who were very close to the accused, 'he did so with the understanding that certain tight affective relations must be respected as they are strongly tied to the right to intimacy/privacy' (G. 291. XLIII: 48).[38] That is, other people can be part of, or at least intimately entwined with, the person. Justices Lorenzetti and Zaffaroni note the historical circumstances of the crime and the time it had taken the state to confront and provide a solution to the puzzle that is Guillermo's identity. In the extended period of the appropriation, 'a whole part of a life has passed where different and decisive aspects of the person's existence were defined and different relationships have been established and maintained and they [these ties] all have very diverse affective and juridical relevance' (G. 291. XLIII: 63). In other words the Justices recognize, at least in part, that when Guillermo asks the Court not to perform the shed-DNA identity test, he is asking that it refrain from upsetting his affective world, or more precisely, stop interfering in his self.

A somewhat similar view of these two worldviews can be gained by examining the way each family – the genetic and the appropriating – views Guillermo. For the family that raised him, Guillermo is their son. While there are many indications in the ruling that their claim is false (for example, the couple's refusal to have Guillermo's blood tested when he was a child and the questionable origins of the documents he holds), their position cannot be disregarded. That is, the fact that there might not be any genetic relation between Guillermo and his appropriators does not imply much within this worldview. While he may not be their biological kin he was raised by them and is socially and culturally recognized as their child. The presumed biological family holds a radically different view. For them, Guillermo is the biological son of the disappeared couple and is genetically tied to their kin group. Based on the results of the DNA test they may, for example, claim that his appearance is similar to that of his genetic parents and that his gestures and likes and dislikes are identical. In fact, in some cases public statements about the physical (and sometimes ideological) affinities

[37] While the Justices recognize these social ties, they do not seem to go as far as Ingold's view (2006a) of the person as a process of becoming through and by way of relationships with others and with the environment.

[38] This was in fact the reason why in an earlier Supreme Court case the Court ruled against forced blood tests.

between the individual who had been identified and his or her disappeared parents emerge following the public announcement of the DNA-test results and the first encounter with the biological family. Noting similarities between the appropriated individual and the genetic family is a productive way to unfold the relations that, until the test had been carried out, had been 'coded' solely in the DNA.

The positions of the two families outlined here are referencing the same person. Each family, however, 'sees' a different person, and from their perspective, these two persons cannot co-exist: Guillermo can either be viewed through his genetic ties or through his social relations (the identity he has held until the DNA test). Within this ontology, he cannot be both. In order for these two different ways of perceiving the person to be sustained, the Court must keep the social separate from the biological. The Justices do so by recognizing or 'seeing' the person from the perspective of his presumed biological family, on the one hand, while also 'seeing' Guillermo through his own eyes as the son of the couple who raised him.[39] These two worldviews are two sides of our western ontology that sustains the distance between organism and person, between nature and culture, or between persons and things.

But the Court does not restrict itself to this position; rather it enfolds these two positions into one, creating a new and altogether different perspective. This new perspective, or what I understand to be a new ontology, collapses the distinctions between the social and the genetic being. Consequently, the person appears to be both at the same time. The clearest articulation of this position is evident in the partial dissent of Justices Lorenzetti and Zafarroni. They argue that the DNA identity test should be conducted on shed-DNA, but that the information should be kept from Guillermo, the direct victim in the case. For them, it is enough that the biological family is informed of his identity. In this way, the family's search can be declared successful and the crime halted. But where does this leave Guillermo? The decision does not erase the point of view that identifies him through his biological identity, nor does it eliminate the point of view that sees him through his social ties. Instead, these two points of view are brought together in one body, thus producing a different kind

[39] The Court does not give any direct indication about the Prieto–Gualtieri couple's position. We can only read and assume their perspective through the Justices' descriptions of Guillermo's presentation of his appeal.

of human being whose perspective – and, in consequence, whose ontology – is radically different.

CONCLUSIONS: THE 'MAGIC EYE'

What would this ontologically different person look like? He would not in fact look any different from others. As Henare *et al.* (2007) note, 'the point about different worlds [different ontologies] is that they cannot be "seen" in a visualist sense. They are, as it were, *a-visible*' (2007: 14, original emphasis). Following from this, it could be argued that the person in the alternative ontology produced by the Court is *a-visible*. Let me try to explain this statement by going back to the idea of perspectivism advanced by Viveiros de Castro (1998) and specifically to his claim that the perspective we hold is assigned to us by our body. In the modern western world we are immersed in the perspective that distinguishes between persons and things (the body and its genetic material). This way, we see Guillermo either as the biological son of a disappeared couple or as the social son of those he calls his parents. This is also why it is difficult for us to 'see' a world where these distinctions collapse. But in the ruling on the shed-DNA test, the Court has produced a different body, which is both social and biological. This new body creates a new perspective or a new ontology. I want to stress that it was not the appellant who had chosen to change perspectives; rather, it was the Court that fabricated this new body and located the individual within this different ontology.[40]

In Argentina, those caught in this ontology prefer not to talk to the media; a few, however, have revealed their world, and in describing their identity restitution they present us with fragments of this alternative ontological position. The case of Carlos Rodolfo de Luccia is a pertinent example. Interviewed by the journalist Analía Argento (2009), Carlos Rodolfo begins to tell his story by describing the moment he was first told of his genetic identity and of his biological family's strong desire to meet him. Though at first refusing, he ends up meeting them briefly. Following this first encounter he maintained relations with both families. When the couple who had raised him was jailed on account of their involvement in his abduction and the falsification of

[40] This may be somewhat analogous to Viveiros de Castro's intriguing analysis of metamorphosis, where he argues that 'metamorphosis occurs at the meeting point of two perspectives, as Guédon observed . . . transformation is not a process but a relation. Nothing "happened" but everything has changed' (2012: 147).

his identity documents, he visited them weekly, and upon their release he spent time in their home. He also met regularly with his biological grandmother and aunt as well as other family members and friends of his biological parents. From the day that the DNA test showed him to be the biological child of Julio and Yolanda until the day he decided to learn more about his personal history, ten years had passed. But knowing more does not mean choosing between two perspectives, so when Carlos got married he invited both his biological grandmother and those he calls 'the family that has raised me' [*mi familia de crianza*] to the event. And later, to commemorate the thirty years of his biological parents' forced disappearance, Carlos organized a religious mass for the disappeared. In concluding her depiction of the interview, Argento provides a transcription of a note Carlos had sent her, which he had asked to append to his life story. There he writes that 'not all histories are the same. They can have the same origins, a birth in captivity [when the mother gave birth while captive in one of the clandestine camps of the dictatorship], following a [forced] disappearance, which is terrible and I think it is imperative to know the truth and to know what happened so it will never happen again' (Argento 2009: 66). And while he is well aware that what his appropriators had done was wrong, he cannot change his feelings for them. At the same time, since he has met his biological family, he has learned to care for them and for his 'parents' [*a mis papas*] (the term he uses to describe his genetic parents). He continues by explaining that while not everyone would understand, 'in my case, I feel that my family grew [was enlarged]' (Argento 2009: 66). And he ends the note by stating that he wants his daughters to grow up with the truth.

Carlos's story is about a sincere and very complicated attempt at keeping the perspective (or ontology) present without reverting to one or another position within western ontology (having to choose between biological kin or social ties). It reflects an attempt to describe what the world is like from within that alternative ontology. The individuals who are caught in this perspective see a different world; for them, the distinctions produced within western ontology between the biological and the social cannot exist. But not many can sustain this ontological position for long. Some attempt to hold to these two ontologies by moving between them, but many revert to the ontological view that is premised on the distinction between the social and the biological.

In most publicly known cases, this implies recognizing the biological identity as the core element of an individual's personhood.

Here comments by individuals who were located and identified using their DNA provide a good illustration of this position. For example, Martín Amarilla who, after the DNA test, found out that he had three brothers and a large biological family, describes his meeting with them in an article in the newspaper *Página 12* in this way: 'The similarity between me and my brothers is very strange. It happened with the youngest, I looked at him and I kept on looking because I had never seen a person so similar to me ... one of them said that there was no need for a DNA [test] because he recognized ... [it was] the ears ...' (*Página 12*, 8 November 2009). Here, an external similarity produces relations. Horacio Pietragalla Corti, another individual who was restituted as a young adult, mentions that after his DNA test he examined his parents' pictures and verified similarities between himself and the photographic image he had of them (Abuelas de Plaza de Mayo 2012). Carlos, whose story was presented above, states in his interview that when he saw his biological father's picture it had a great impact on him. 'It was like seeing me with a hairdo from that period. It is my face, it is shocking' (Argento 2009: 50). And, Juan Cabandié Alfonsín, restituted in 2004 as a young adult, mentions how when he had doubts about his identity he talked to his sister, and they both agreed they did not look alike; the DNA test confirmed that he was not genetically related to her (*Página 12*, 28 March 2004).

Most often in these cases, genetic identity is recognized as the core element of individual personhood, and once the choice is made the individual cannot go back. A powerful indication of this move can be read in an open letter written by restituted individuals and their siblings. The letter was made public following controversial declarations by perpetrators of the crime of appropriation. They write: 'we can say that we recuperated our true identity and that has allowed us to be ourselves and not that which others have wanted us to be, first assassinating our parents and then changing our identity ... we are today free because we can choose what to do with our history ... we are free because we recuperated that which was stolen from us, because we put things in their place and this way we could understand our DNA, that which no one can change' (posted on *Abuelas* website). To tell their story of identification and restitution by stating that they now 'understand' their DNA is, to my mind, to take on a perspective that takes the distinction between the social and the biological completely for granted.

Once the choice is made and the individual has accepted his or her new identity through acts such as changing a name or working toward creating affective ties with the biological family, these individuals cannot go back. By taking on what the Court defines as their 'biological identity', these individuals are once again located within the prevailing and powerful ontology of the West. The story told in the above letter is the public face of the struggle for justice and identification in Argentina. This is because the ontology that brings together persons and things is hard to describe, particularly to others who are caught in a perspective that strongly maintains the distinction between them. Guillermo Gabriel Prieto presents a particularly significant case to ponder, because although he had chosen his social family over his biological identity, he was made to enter into an alternative ontology through the Court's decision to allow the shed-DNA identity test. What would it take, then, to be able to be in this alternative ontology? Or, at the very least, to recognize its parallel existence?

One way to think about these two ontologies and their co-existence is to consider the work we do when we look at a picture of the 'Magic Eye'. When we first examine a 'Magic Eye' picture, all we see is a random pattern; but as we learn to 'relax' our eyes, that is, to change our mode of perception, a different picture emerges (sometimes in 3D). This one is very different from the original pattern. While both pictures exist on the page, to be able to see them both we must change our mode of perceiving the world. This alternative way of looking is, I believe, becoming increasingly significant for our existence in a world that is rapidly changing in the footsteps of new technological and scientific advances. We would do well to exercise and develop it further.

BARBARA GÖTSCH

7

Reflections on a collective brain at work: one week in the working life of an NGO team in urban Morocco

The following scene is set in Rabat's district of Agdal; a neighbourhood which boasts many office buildings, upscale apartment blocks, a fancy shopping boulevard, as well as the university. We will enter the office of a small non-governmental organization, located on the third floor of a white modern building, just off the main boulevard.

A sunny Thursday morning in October 2007, shortly after the end of Ramadan, the whole team was present at the office: eight people squeezed into Naima and Thérèse's office room, which simultaneously served as the office kitchen. Brahim brought folding chairs so that most team-members found seats; others, like me, stood at the back near the door. Thérèse – and her computer – were situated at the opposite end, for everyone to have a good view of the screen and the PowerPoint presentation to be projected on it. Ahmed, the team leader who had ushered everybody in, excitedly explained that it was extremely important that this slideshow should have a positive impact. The *diapos* should reflect the knowledge and expertise of the whole team. '*Je veux profiter de toute votre intelligence*', he said. Thérèse started to read out loud, facing the computer while reading, turning around at regular intervals to see the reaction of the team. They went over every sentence, word by word, examining whether they had chosen the right expression for the purpose and for the expected audience. Not everyone was as enthusiastic about this ad hoc meeting. Naima had seen most of it already, as she had sat in on several collaborative work sessions where they had jointly devised the content. She was slightly irritated that, again, it was Thérèse who was writing – who was 'in

Biosocial Becomings: Integrating Social and Biological Anthropology, eds T. Ingold and G. Palsson. Published by Cambridge University Press.

control' – and that Thérèse had apparently added some phrases of her own, probably at home, without consulting the others. Meriem – the new *collaboratrice* – looked very concentrated and slightly puzzled. She made an effort to understand what it was all about. The technical terms were one thing, but the acronyms of the projects and project partners, as well as the stories surrounding the people mentioned, were a completely different matter. She didn't raise her voice, tried to take it all in, hoping that everything would become clear in due course. Ahmed sucked up all the commentaries, and encouraged the colleagues to reflect, to think strategically, to imagine creatively how these phrases might be received by future interlocutors. Brahim, resourceful as ever, got into full swing, suggesting amendments, putting himself in the position of the future audience, and enacting pieces of pretend play in which he performed several imagined roles in the future encounter. Others joined in and there was laughter. We all felt involved. A momentum had arisen. After about an hour it was over. The room was filled with hot air; one could hardly breathe. We gradually removed the chairs and left the room, rosy-cheeked and exhausted.

This scene is based on my experience as an ethnographic fieldworker. It took place in the autumn of 2007, five months into my participant observation with a team of co-workers for my doctoral thesis in social anthropology.[41] The team comprised roughly seven persons, all co-workers in a non-profit organization. This constellation of co-workers had existed for the most part for at least 10 years. To speak with Ingold, it was a group of 'wayfarers' who had been 'inhabiting' the same workspace and had spent much of their spare time together for quite a while (Ingold 2011: 12). Except for the team assistant, all *collaborateurs* and *collaboratrices* were trained psychologists or pedagogues, who worked as trainers, researchers and lobbyists on issues of children's care and education. To preserve the organization's anonymity I refer to it by the pseudonym *Education Daba*, which, in a mixture of French (or English) and Moroccan Arabic, means 'Education Now'.

I had started fieldwork with this group of people in May 2007. In my second week, I witnessed increased anxiety among the team members, as a visit by representatives of an international consultancy firm was announced. The consultancy firm, acting on behalf of a prestigious Moroccan royal foundation, was to fathom the team's competence and

[41] Research on my PhD is still ongoing. The insights I share in this chapter are therefore still preliminary.

potential willingness to contribute to a large-scale education programme that the foundation sought to establish. The visit went very well, as far as the team could tell, and so they were expecting to embark on a new challenge. But then, nothing further was heard from the consultants for the whole summer. Rather unexpectedly, at the beginning of October, the team was contacted again and asked – within the frame of two weeks – to present an outline for this potential programme. For a fortnight, the team was caught in a frenzy of activity. In this situation Ahmed, the team leader, repeated something he had hinted at before, but which he thought particularly important under the current circumstances: he wished the team to pool their energies and to work as *un cerveau ensemble*, that is, to work together as 'one brain'. The scene I described above took place on the day before that on which the decisive project presentation meeting was scheduled, and it represented the culmination of their joint preparation efforts. During this one hour, those present indeed experienced the pooling of energies; there was a high degree of intersubjectivity. For the team, the effort was worth it; after all, much was at stake. Were the outline approved, the chances were high that the team would be invited to fine-tune the programme and then do the actual teaching of large parts of it. For the co-workers this meant an interesting teaching activity ahead, the potential consolidation of their good reputation – 'sealed' by the *Makhzen*, the royal house and its apparatus – as well as large-scale funding for a kind of training they thought long overdue for the country.

But what exactly happened there and then? How was this experience of a 'collective brain' to be explained? I had arrived in the field with an overall research interest in cognition, communication and cooperation. This situation, then, presented me with a puzzle, to which I had to give serious thought; a puzzle, though, with which I am still grappling. In what follows, I shall endeavour to reach a better understanding of this intense situation of collaboration. I will therefore reflect on work practice in the team, including a description of the goings-on in the week leading up to the important meeting and, of course, an examination of the rehearsal situation itself. This reflection will also include some thoughts on my own experience of growing attuned to this group of people, since this may help put the pieces of the puzzle together.

I will approach this exercise with the help of several strands of literature. I begin by drawing on the works of Lave and Wenger (Lave 1988, Lave and Wenger 1991, Wenger 1998), with their focus on social learning in so-called 'communities of practice'. Turning to the literature on human sociality, I draw on approaches in cognitive and linguistic anthropology, sympathetic to developmental systems theory, which seek to integrate the social and the biological by highlighting the processual character of human life. However I have also been attracted by a further strand of literature, emphasizing the biological and psychological aspects of sociality, looked at from an evolutionary perspective (in Enfield and Levinson 2006; see also the collections edited by Sussman and Chapman 2004, and Goody 1995), which accords a central place to face-to-face interaction in accounting for human socio-cognitive abilities. The majority of these scholars argue that our unique kind of social interaction is at the root of what makes us human.[42] However, while acknowledging that we cannot ignore evolution when thinking about social life, I do not agree with the stark contrast overwhelmingly drawn by these authors between universal human capacities and culture-specific knowledge, nor do I share their preoccupation with the evolutionary roots of sociality. For my part, I am more interested in the iterative process, the day-to-day growth of sociality, seen as a biosocial ability that is developed in social interactions among people.

A 'COMMUNITIES OF PRACTICE' PERSPECTIVE

The 'communities of practice' approach goes back to the work of Scribner and Cole (1981) and Rogoff and Lave (1984), but was primarily developed by Lave and Wenger (Lave 1988, Lave and Wenger 1991, Chaiklin and Lave 1996, Wenger 1998) during the early 1990s. Wenger developed the idea further in his *Communities of Practice* (1998), where he portrayed processes of learning and situated cognition among medical claims processors in a big US insurance company. In essence, the approach proposes that people who engage in joint activities learn by way of this engagement in the contexts of their social interaction, whether as members of professional work-groups, street gangs, families, associations of hobbyists, or groups of online gamers dispersed across several countries. Wenger spells out three dimensions that are essential for a social grouping to qualify as a community of

[42] Levinson proposes that 'the roots of human sociality lie in a special capacity for social interaction' (Levinson 2006: 39), which he terms the 'human interaction engine', i.e. 'a set of cognitive abilities and behavioral dispositions that synergistically work together to endow human face-to-face interaction with certain special qualities' (2006:44). He sees this 'interaction engine' as establishing the 'generic constructional principles on which cultural diversity may be built' (Enfield and Levinson 2006: 13).

practice: mutual engagement, a joint enterprise and a shared repertoire (Wenger 1998: 73).

Though it has its roots in the theory of practice as developed by such diverse thinkers as Marx, Bourdieu, Sahlins and Giddens, the community of practice approach is specifically interested in the ways in which practice intersects with issues of everyday situated experience and identity, through the process of learning. Cognition, in this perspective, is an intrinsically social phenomenon. As Lave puts it, "cognition" observed in everyday practice is distributed – stretched over, not divided among – mind, body, activity and culturally organized settings (which include other actors)' (1988: 1). This perspective on learning in communities of practice converges with Ingold's (2000) more encompassing approach to the development of skill, in that both stress the inherently social character and situatedness of learning. Ingold defines skills as 'the capabilities of action and perception of the whole organic being (indissolubly mind and body) situated in a richly structured environment' (2000: 5). And he insists that skills are 'not transmitted from generation to generation but are regrown in each, incorporated into the *modus operandi* of the developing human organism through training and experience in the performance of particular tasks'. In this process of enskilment, novices, newcomers or apprentices go through a phase of what Lave and Wenger (1991) call 'legitimate peripheral participation'. This is a time when they are entitled to hang around 'experts' but are not expected to perform at the same level of expertise as more experienced members of the community.

Work practice in 'my' team involved two major activities: individual study and writing activity and verbal exchange on organizational matters as well as on specific subjects, which at times also resulted in joint writing processes. Overall, there were many lengthy discussion meetings and collaborative work sessions. Legitimate peripheral participation thus involved a lot of sitting in on meetings. For newcomers this meant sitting in without talking – to the extent that it was basically expected of them to ignore the team leader's frequent invitations to come up with ideas or comments. Brahim, one of the senior staff, told me that when he joined the *équipe* some ten years ago, he had sat in for six months, just listening to what the others had to say and only speaking up when addressed. During my time with them, I observed him as one of the most prominent and resourceful speakers in every meeting. It was the same for Thérèse, who, I was told, had appeared rather insecure at the beginning, but through sitting in and listening, had gradually acquired legitimacy and practice in speaking out loud. She, too, became one of the most prominent and eloquent speakers in the group. As for myself, the fieldworker and newcomer, I seem to have undergone the same phase of silence. Only after about four months of participant observation do I find my own words quoted in the fieldnotes which capture team-meetings.[43]

It so happened that at the beginning of October, just before the daunting project request came in, the team had hired a new *collaboratrice*. The term *collaboratrice* was deliberately used in reference to the position and later on to the actual person. It carried connotations of competence and equality compared with those on the team who also saw themselves as *collaborateurs* or *collaboratrices*. The equally newly hired administrative aid was not called *collaboratrice* but *assistante*, and one of the junior staff members, who could not boast high academic credentials ('merely' a bachelor's degree), was only occasionally referred to as *collaboratrice*. Clearly, the questions at stake in the particular practice of this community were not merely of competence; they were also of status, linked to the linguistic, academic, religious and socio-economic backgrounds of the candidates, as well as to gender and age.[44] At any rate, Meriem, the incoming *collaboratrice*, immediately understood that it was expected of her to undergo a phase of silence so that she could then acquire the status of full participant.

Thus, a fully legitimate participant knew when and how to talk. How to talk was also a question of how to present a story,[45] whom to refer to, how to involve those present in the story-telling, what conclusions to draw from events, and how to approach problem-solving. Furthermore, full participants were competent in presenting themselves and in displaying their knowledge to others, and showed an

[43] Of course, I interacted verbally with them right from the start. It was only in the forum of the team meeting where I felt I had to stay silent for many weeks. Scheduled team-meetings in this group, albeit internal, had an official character. One could characterize them as a type of front-stage performance in Goffman's (1959) terms, whereas talk in the corridor or in the converted kitchen could be characterized as back-stage talk. The scene I recounted at the outset is an exception to this rule: here, the kitchen served as the venue for a front-stage performance.

[44] As noted by Tusting (2005), issues of language use and literacy, as well as power relations, figure only marginally in the original 'communities of practice' literature.

[45] This resonates with a study by Cain of Alcoholics Anonymous, reported in Lave and Wenger (1991: 79–84). Cain found that to be eventually accepted as a 'non-drinking alcoholic' in the group was connected to a specific way of presenting oneself and telling stories.

interest in actively participating in endless meetings as well as in developing knowledge and skills in the valued practice of joint social reasoning and perspective taking.

In almost all meetings and cooperative work sessions, the team self-consciously reflected its positioning vis-à-vis other stakeholders in the domain. When it came to getting ready for any meeting with powerful people from the outside – be they representatives of funding agencies, ministries, or international organizations – they not only stood back to reflect but also creatively and playfully prepared for these encounters. This was so important to them because to see something that at least resembled their vision of the future of education in Morocco realized, they needed support from many sides: from those in power as well as from those who were to realize projects and policies on the ground, and of course they needed funding for activities. It was not so clear to them, however, who had a say on diverse issues, what these actors' positions on educational matters were, and how the power-game between them was being played. The political landscape under the new king was in transition.[46] There were ministries (of youth and sports, family, religious affairs, etc.) with partly diverging agendas on matters of education, as well as diverse state agencies, committees, royal commissions and foundations, which all proclaimed that they were acting somehow or other in the interests of the future of Morocco's children. The team realized that to be successful one needed to know what these institutions were aiming at, and to design propositions accordingly. Also, and very importantly, they were acutely aware of the fact that they needed to know their respective interlocutors in person, first, to establish a certain rapport, and second, to be able to anticipate what these people might expect of them.

As a result, preparation sessions for important meetings entailed a lot of perspective taking, hypothesizing about the thoughts and feelings of these others and imagining future scenarios to the extent of engaging in pretend play. By listening in on many, many occasions and gradually trying out some hypothesizing themselves, newcomers were

[46] King Mohammed VI succeeded his father to the throne in summer 1999. In his first decade as a monarch he gradually introduced a rhetoric and politics of moderation, changing much of the personnel in the politically relevant elite (Perthes 2004) and inviting the participation of civil society organizations. In 2011, as a reaction to the so-called 'Arab spring', he had the constitution changed, granting parliament more power. The political system, however, is still based on patronage, with the king at the centre and all political power firmly in his hands.

linguistically socialized (Schieffelin and Ochs 1986) into this habit of perspective taking *à voix haute* (aloud). Hence full participants boasted high competence in at least third-order intentionality or higher. This again could only be effective in combination with knowledge of the socio-political environment in which the NGO was operating. Thus, team-members were expected to stay updated on political events, and encouraged to invest in contacts with people from within the NGO scene, with the educational sector, and, if possible, with people linked to the Makhzen.

Very importantly also, the 'how' involved near native competence in French. After all, this group was a distinctly French speaking community of practice, especially so far as literacy was concerned. Competence in Modern Standard Arabic, the country's official language (besides Amazigh, introduced as the second official language in 2011), was only of secondary importance for their work. This was partly due to the fact that most senior colleagues were not fluent enough in Modern Standard Arabic to read, write and speak freely in it. More important was a certain level of bilingualism that included the local Moroccan Arabic dialect, for which there is no written standard. As a result, social or situated learning involved learning how to express oneself appropriately in French. The senior staff members were all educated in French, the juniors less so. It hardly needs saying at this point that the seniors automatically drew their theoretical impulses from specialized literature in French – and expected juniors to do the same. Unsurprisingly, conflicts often revolved around questions of language use. This issue was aggravated through the fact that team members were expected gradually to specialize in a specific subject area, so as to be able to contribute knowledge others did not possess to the same extent, and so as to then qualify as trainers and eventually as authors of educational handbooks. Once you were entitled to work as a trainer and were even commissioned to write an educational handbook, you had definitely passed the threshold of full participation. These handbooks were written in French and later translated into Arabic. Clearly, power relations in this community of practice were played out through language use. Training was also preferably given in French; a fact that did less and less justice to the social and political circumstances in the country, where the introduction of a politics of arabization in public schools had meant a gradual weakening of fluency in French in the overall population.

As for my role on the team, officially, and especially for everyone coming from outside, I filled the position of *stagiaire* (intern); unofficially, though, they all knew and accepted that I was doing ethnographic fieldwork. The first few months were mainly spent running errands and making photocopies but otherwise being allowed to just 'hang around', doing participant observation. After a few weeks, I was treated by the seniors (the dominant group) as naturally belonging to the juniors, and I was accepted by the latter as if I were a 'normal' colleague, with conversations gradually gaining in depth as trust grew. We juniors, thus, spent most lunch breaks in the office, eating precooked food or buying sandwiches from the street, whereas the seniors mostly went home where they had a meal prepared for them by some household help. I was also included in several semi-official off-site activities, such as lunches in nearby restaurants, the farewell party for a retiring senior colleague, participation in a conference and, much later, a one-week trip to Taroudant, a town in the south of Morocco, for a training course. I particularly befriended one junior and one senior colleague, who then frequently took me for shopping sprees in fancy Agdal, walks in the old medina and visits to the beach, and invited me to family gatherings and other 'spare-time' activities. By the end of my sojourn, I had visited the homes of almost all team members at least once.

In my fourth month, Naima and Thérèse, two seniors, asked whether I would like to contribute a small input of my own to a training event. They were working on the topic of 'diversity' in educational contexts in one of their training workshops, and were hoping I could say something on the topic of 'identity' and ideally combine this with some interactive activity. I gladly accepted, grateful to be able to give something in return. This also involved my being invited to contribute actively to collaborative work sessions in which the one-day training event was prepared. In a way I became a member of a new group – that of the core training preparation team – in which I was expected to make suggestions related to diverse issues and also to express my opinion on the suggestions provided by others. Also, we shared insights gained from specialized literature in French, respectively in English and German. In other words, we engaged intensely with each other in a way which for me, the apprentice, facilitated an enormous growth in enskilment in this community's practice. Moreover, through spending time with each other during preparation, focusing on a specific task, we gradually gained some common ground, both on the topic of education and on what counted as problematic, important or unimportant in our work. This not only allowed us to dispose of a great amount of common knowledge; it also intensified mutual trust. And that, in turn, laid the ground for many of the joint activities that lay ahead, including the more egotistical endeavours of my recording interviews and conversations, as well as my involvement in the scene I recounted at the beginning of this chapter. When I finally appeared in the training event, I felt very stressed; my heart was beating wildly and my hands were sweating at the sight of twenty practitioners in education, all far older than me, who expected me to say something meaningful to them. One could say that I had to get my sea legs, as Palsson (1994) would have it. His example is of Icelandic fishermen, who refer to the recovery from seasickness as an important milestone in becoming skilful in seafaring. In other words, I had to overcome a social situation which was physically straining for me. I hadn't completely found my sea legs by then, but I certainly had come a long way that week.

After my involvement in planning the training event as well as my short appearance as a trainer, I was treated differently by the seniors. I felt that my status was upgraded from legitimate peripheral participant to candidate for full participation. As a result, they started to compare my performance to that of other juniors including the newly arrived Meriem, who hadn't quite found her feet in the team yet. Although it was clear that I would leave in a few months' time, I was treated as somebody who had the potential to become a *collaboratrice*, and thus as someone who was potentially favoured over other junior candidates. Inadvertently, for the other juniors I had become a rival for attention and appreciation from the seniors. Although I felt I belonged more – and had made an important step in my research process as an ethnographer – I also felt uneasy, as I entered a liminal in-between status, both in time but also in terms of social relations.

THE WEEK LEADING UP TO THE IMPORTANT PRESENTATION MEETING

Let me return to the team's preparation efforts for the project presentation meeting. They were to present their suggestions to an audience comprised of members of the consultancy firm as well as to representatives of the royal foundation, on a Friday morning in October. In the week leading up to the decisive event, Monday and Thursday were especially busy preparation days, as the other days were occupied with other engagements. Without further ado, the Monday meeting, or weekly *réunion de synthèse*, was extended and turned into a whole day of sitting together, exchanging ideas. The co-workers did not have to start from scratch. First, they had the original rough outline which they had designed in May; second, some colleagues had worked on elaborating this outline, notably Thérèse and Brahim, two senior members of staff.

Early on in the Monday meeting marathon, a dilemma emerged which the team had difficulty coming to terms with. They had to resolve the tension between what they thought was expected of them in the meeting – of which they were not completely sure – and how much of their own expertise they wanted to give away at this point. The full extent of the dilemma unfolded in the course of the conversation. At first the team leader only urged the others not to go too fast in elaborating the project content. 'We should not go faster than the others, because this repeatedly has led us onto the wrong track'. Later on he warns, 'Let's wait how they evolve with us... possibly this will end in smoke'. Brahim, who had just introduced the idea of structuring teaching subjects in the form of larger units, argues that through this 'we will have a very clear idea of what we want to do without going into details'. Thérèse does not feel they should slow down. She thinks that they should know as much as possible about what they eventually want to do as she is worried about the quality of the product they are to present. She says, 'If we want to give these people the document on Friday, this is going to be very complicated because there shouldn't be too many errors in the outline, yet we have very little time to develop the programme content in depth'. Also, she worries about who could teach what. This is something Ahmed takes seriously, too; he doesn't want the programme to be in contradiction with the direction that the team, and thus potential teachers, may take in the future. At some point Ahmed warns, '*daba, c'est gratuit*', now our work is for free. Payment would come in only if they were granted the project, which was not at all certain. Several other stakeholders in the sector of education were similarly approached by the foundation. It was not clear to the team which role these other stakeholders played at present, and what this could mean for the future. At worst, somebody else could use their expertise to realize the project and could subsequently be associated with it, both publicly and, very importantly, within the royal arena. This would pose a severe threat to the public image of the organization, and put some senior colleagues' ambitions for upward social mobility at risk. In short, there was a lot of apprehension concerning the future – a feeling of apprehension, though, which drew them closer together. In the process, the team members expressed confidence that what they were devising was a great project and that they, if anyone, were the right people to realize it – which further boosted their desire to put forward something really good.

Faced with this dilemma, the day was spent partly in advancing the programme, and partly discussing what could be expected of them and how much content they should present on Friday, considering the above-mentioned uncertainties. Gradually, they developed a joint vision of what the 'others' could expect of them, what the event should look like, and how much content they wanted to give away at this point. In an essay on the joint telling of stories in American middle-class families, Ochs, Smith and Taylor (1996) called a similar phenomenon the telling of 'detective stories'. This term was chosen to emphasize the puzzling qualities of certain states of affairs and the effort to come to terms with them through co-narration. Through the participation of others present, who ask questions and voice suggestions, a story or joint vision eventually emerges. According to Ochs and her colleagues, this represents a 'socially accomplished cognitive activity' (1996: 95). In their words, detective stories

> grow out of the process of grappling with incomplete understandings of life's complexity and multiple 'truths' and perspectives, often manifest (and in conflict) in everyday events and their telling. Initial narrators often seek the kind of co-narration that helps further their own comprehension of their stories and gives meaning to their stories and their lives. (Ochs *et al.* 1996: 99)

Thus, in a way, one could say that the team were puzzling over a detective story, jointly constructing a vision of what challenges lay ahead and how to face them, solving their problem through co-narration.

Looked at from the communities of practice perspective, what the team engaged in were processes of the negotiation of meaning and reification. Wenger (1998) uses 'negotiation of meaning' in a very broad sense, defining it as 'the process by which we experience the world and our engagement in it as meaningful' (1998: 53). He deliberately conceptualizes it broadly in order to capture two different aspects: the idea of reaching an agreement through negotiation, and the idea of an accomplishment, achieved through gradual refinement. According to Wenger, negotiation of meaning in communities of practice conveys 'a flavour of continuous interaction, of gradual achievement, and of give and take' (1998: 53). He sees participation and reification as constitutive of processes of negotiation of meaning.

'Reification' stems from Latin and literally means 'to make something into a thing'.[47] In preparing for the project presentation, the

[47] As with 'negotiation of meaning', Wenger uses 'reification' in a very broad sense. He acknowledges though that his usage is close to Lukács's treatment of Marx's

team were engaged in creating a specific reification of their ideas, a slideshow presentation. But in the preparation process they were involved in other types of reifying, too. They constantly produced discussion notes, drew pictures to illustrate their ideas, and amended older versions of the project outlines. Also, while they were working on the presentation, they compared what they were developing at present to the contents of a series of educational handbooks which they were in the process of publishing. These handbooks represented already materially manifest reifications of the group's work and were ever-present in their thoughts and discussions. The books had enormous importance, almost gaining a life of their own. When discussing the structure of the course outline, for example, it was difficult for the team to imagine a structure that would not follow the division of the series according to its book titles. Starting out from the importance Wenger (1998) assigns to reification, Barton and Hamilton (2005: 14–35) point out that it is important to look at literacy in communities of practice. In their view, what the team were engaging in would have been a 'literacy event' – that is, an event in which interaction occurs in the presence of texts and in which a good deal of interaction refers to them. These mediating texts, in turn, are seen as changing what is said and how it is said (Barton and Hamilton 2005: 17ff.). As pointed out earlier, literacy practices in this community reflected power relations and resulted in an overall dominance of French, both in conversations and in written texts.

The other element that Wenger (1998) identifies as a constituent in processes of negotiating meaning is 'participation'. With that he wants to stress the inherently social character of all processes of working, communicating and thinking in communities of practice. In this sense, we are never alone, even if we are physically isolated from others. We always take the (albeit attributed) thoughts and feelings of others with us. For 'my' team, participation had particular importance for another reason as well. The team leader accorded great importance to the active participation of all team-members. This went along with a doctrine of transparency, but also served the purpose of sharing ideas and of creating opportunities for voicing different viewpoints.[48] Also,

idea of a 'process by which societies endow social phenomena with thinglike properties' (Wenger 1998: 287).

[48] The voicing of different viewpoints was certainly limited. This was still Morocco, where a principle of seniority applied – elucidated, for example, in Hammoudi's (1997) *Master and Disciple* – and where especially youngsters and newcomers were expected to display respect and refrain from criticism.

and very importantly, the participation of all team members in as many meetings as possible served pedagogical ends. In these team meetings, participants overtly and covertly were taught – and taught each other – specific ways of thinking, feeling and acting (Ochs 1986: 2). In other words, their attention was directed to specific ways of being in the world (Ingold 2000, 2001b).

THE 'COLLECTIVE BRAIN' REVISITED

Let me return to the ethnographic scene I narrated at the beginning of this chapter. In this scene a group of collaborators gather in a small room (the communal kitchen) in order jointly to go over a computerized slideshow presentation that is to be presented to a highly important potential future client the following day. All of the co-workers have a good view of the screen, one colleague reads out loud the content of each page, the others are encouraged to mentally anticipate the 'real' presentation and suggest amendments. Here and there the team-leader orients the participants' attention to one or the other aspect. Overall, he declares that he wants to benefit from all their intelligence for the presentation to become truly perfect for the occasion. This situation was experienced by those present (including myself) as particularly intense. People commented upon it when leaving the room and it was also visible from their faces.

I discern several socio-cognitive phenomena in this situation, which I will discuss in the following.

First, to my mind, the team members reached a high degree of shared intentionality. The term 'shared intentionality', or 'we-intentionality', is used in psychology to describe collaborative interactions in which several people share psychological states with others (Tomasello and Carpenter 2007: 121). In the case at hand, the team members shared a joint visual perspective, they shared a large amount of knowledge concerning what they saw on the screen (including how this document came into being), they shared a joint goal (producing a PowerPoint presentation to help them make a good impression during the project presentation), and they shared many ideas of how to achieve this goal. In addition, they continued verbally to share further ideas with each other while they were in the room, including joint 'as-if' cognizing when Brahim started to pretend they were in the actual presentation situation.

Second, I would like to take up two expressions which I introduced in the preceding paragraph, and which I deem extremely important for this context: *joint attention* and *common ground*. These two conditions, according to Tomasello and his colleagues from the Max Planck Institute for Evolutionary Anthropology, are the prerequisites for developing shared intentionality in the first place (Tomasello and Rakoczy 2003, Tomasello and Carpenter 2007, Tomasello 2008). In other words, from a perspective in developmental psychology, the ability to direct our attention to the same object as another person, and be mutually aware of this joint attention, is a prerequisite for later being able to direct our thinking or feeling to the same idea as another person. During preparation time, the team members had invested in joint attention to each other's behaviour, to each other's verbal expressions, and to artefacts that reified their work. Joint attention in collaborative activities not only means that several people focus on the same issue at the same time; they are also mutually aware of the fact that they share this experience. This again facilitates intersubjective understanding. I think shared attention is something we humans are very good at. We have learnt to 'tune in' to the world of others, as well as to get others to be 'tuned in' to our world (Tomasello and Rakoczy 2003: 125). In the words of Tomasello and Rakoczy, it is the understanding of persons as 'intentional agents who have a perspective on the world that can be followed into, directed, and shared' (Tomasello and Rakoczy 2003: 125) which is at issue.

Common ground – a concept closely associated with the work of Clark (Clark and Brennan 1991) – essentially represents any kind of knowledge that people share and know that they share. The NGO team, as a bunch of wayfarers, had built up a base of common ground by virtue of their longstanding experience of knowing each other and having worked together, which was then intensified by jointly collaborating on a specific task. Tomasello argues that collaborative work forms a special case of joint attention, leading to a particularly strong common ground, when he says that 'common ground created by top-down processes in an immediately copresent perceptual environment – specifically in the joint attention of collaborative activities – is in some sense primary' (2008: 78ff.). In addition, Ahmed, the team leader, and Brahim, a senior colleague, constantly engaged in further 'grounding' as they kept informing their colleagues, who were less knowledgeable or experienced in the ways of the world, about whatever they thought relevant about the future encounter. Both had had longstanding experience with different ministries, government agencies and royal foundations, and prided themselves on knowing *comment eux fonctionnent* (that is, both how these bodies went about their work and what their mind-set was like). None of them had had any experience with – albeit Moroccan – representatives of an international consultancy firm, though. Speculations about the future encounter were therefore strongly based on the impression the consultants had made on them in previous encounters and on what they knew about the royal foundation for which these consultants were working. This, together with the longstanding experience they had had with similar 'types', still made Ahmed and Brahim competent advisors for the rest of the group. So when they went over the PowerPoint presentation together, Ahmed and Brahim would regularly discuss their colleagues' suggestions for amendment in the light of their own vision of the event.

Third, it was arguably not 'merely' the sharing of attention, and in extremis the sharing of mind states, which was striven for. Ahmed explicitly stated that he wanted to 'profit from all their intelligence' and that the presentation should reflect all their knowledge and expertise. This means that he was well aware that there were different viewpoints as well as different kinds of expertise present. He wanted the team members to bring all of their perspectives together. What the team leader was looking for, and what conceivably happened, resembles what Hutchins has called 'socially distributed cognition' (Hutchins 1995, 2006). Hutchins's example is the close cooperation of the people who navigate a huge navy ship. The distributed cognition perspective highlights the convergence of different skills and knowledge, enacted by different people simultaneously in performing a specific task and working towards a joint goal. Summarizing his insights, he concludes that 'the social distribution of cognitive labour increases the variability in the choice space. The conversational practice of taking turns suggesting and evaluating options creates a cognitive system that is likely to explore a wider range of alternatives than would be explored by any navigator alone' (Hutchins 2006: 391). In this light, common ground not only allowed the colleagues to have a similar base of knowledge and to come to similar conclusions; it also afforded them the possibility to draw divergent conclusions, which could be discussed in order to enrich the overall picture.

Fourth, trying to consider as many perspectives as possible is a good recipe in preparing for a social situation in which many variables are unknown. This can involve different kinds of knowledge and expertise, but also perspective-taking as part of social reasoning, as well as engaging in pretend play. It may sound rather unusual, but pretend play was practised on a regular basis, casually and in the midst of discussions on various topics, always anticipating some future situation. Brahim, a senior colleague who was particularly gifted at it, also did it in the vignette I presented at the outset. When Brahim starts pretend play, he puts himself in several roles, imagining the mind states of several future participants, including those of his own colleagues at present and in the future. It is an excellent exercise in empathizing and especially in taking perspectives. Moreover it is great fun to watch. In the team, pretend play was almost always accompanied by laughter. And interestingly, laughter creates hormonal reactions (it releases endorphines) that make us happy and create bondedness among those who are co-present, creating a communal experience (Dunbar *et al.* 2012). This may explain why I felt that a momentum had arisen and that people left the room rosy-cheeked. Also, much as for children, pretend play in a creative manner prepared the team for cognizing about agency and social constraints in the world outside the office. It helped them rehearse for critical social situations, playfully taking into account threats and opportunities – as management people would say – in the socio-political environment.

Fifth, looked at from the perspective of developmental psychology, the ability to engage in pretend play, as well as in joint or shared attention, makes possible the gradual development of a 'theory of mind' in children. Here I am introducing yet another psychological term, not as a diversion, but because the socio-cognitive phenomena described above all tie in with 'theory of mind', and because my team of NGO activists actively worked on developing their 'theory of mind' skills further – although they would not call it that. 'Theory of mind' is a psychological term that refers to the ability (in a metaphorical sense) to mind-read other people's thoughts and feelings, to take on other people's perspectives, and ultimately to know that other people act according to *their* beliefs and desires and not according to one's own. It ranges from casual (unconscious) conversational inference to overt social reasoning. Bloch (2007) discusses this socio-cognitive ability in an essay entitled 'Going in and out of each other's bodies'. He claims that the importance of 'theory of mind' for human social cognition cannot be underestimated. The somewhat unusual title of his essay serves to emphasize the way in which humans, sometimes consciously but most often unconsciously, mentally follow into the minds of others. He argues that the boundaries of biological entities are indeterminate and arbitrary, considering the high degree of mental interpenetration.

There is some debate about the extent to which the 'readability' of other minds is subject to cultural variation (Duranti 2006, 2008, Danziger 2006, Robbins and Rumsey 2008). In Pacific societies, for example, there is a widespread belief or doctrine that it is extremely difficult to know what other people think or feel and that therefore one ought to refrain from speculating about it. In Morocco, quite the opposite seems to be the case. According to the Qur'an and the Hadith it is the intention (نيَّة – *neyya*) that counts. God will judge you according to your intentions, not according to your actions. Correspondingly, in Morocco you frequently encounter explicit verbal references to what this or that person might think. Moroccans, if you will, are prepared to talk about inner states,[49] maybe not so much about their own, but hypothesizing about the thoughts and feelings of others is common. As will be clear by now, my NGO team was no exception. Team members endlessly engaged in sessions of social reasoning, where they jointly analysed the social world around them and formed hypotheses as to what certain people might think, feel or expect. I did not ask them about inner states myself until I had listened to them speculating wildly about them among each other, in both front-stage and backstage situations, for months (*pace* Duranti 2008).

As I intimated earlier, the socio-political environment in which this NGO operated was socially complex and full of contestation. They wanted to make a change, to contribute to socio-economic development in the country. In many ways their social imaginary (Appadurai 1996) converged with that propagated by the king: they followed a human rights based approach and welcomed the king's initiative in promoting good governance and good citizenship as well as his moves in creating more gender equality in the country. These were keywords they readily integrated into their project proposals and educational publications. However, the values associated with these keywords were not at all widespread in the country, let alone known, and on the whole they belonged more to the realm of rhetoric than to politics on the ground. To agree with the king was one thing; to translate this into the realization of educational programmes was another. For this to happen, the organization needed the support of lower-level actors, and this was not so easy to obtain. Some of these gatekeepers received them well, others resented them for personal reasons or for what they

[49] See Bloch (2011) for a discussion of the difference between people who have developed a habit of talking about their inner states and those who have not, and what this means for an anthropology of the self.

represented: very well educated, mostly foreign trained members of an educational elite, oriented towards France and 'the West'. And there were others who were opposed outright to what they were doing: be they Muslim groupings that did not want to see any weakening of the importance of Islamic learning in education, or conservative agents from within the school system that did not tolerate any sort of criticism of the existing order. Adversaries of all sorts were potentially dangerous – for the longevity of the organization, but also for the well-being of individual team-members. Therefore, the team had to think ahead. They needed to convince other people while saving their interlocutors' face, often having to make sure that these others did not see that they were being manipulated; and they had to protect themselves. They needed to anticipate future social and communicative situations to accomplish their work and advance in society. After all, as Enfield concludes, 'there is no time out from the social consequences of communicative action' (Enfield 2006: 423). Their strategy of joint social reasoning and the continuous growth of their 'theory of mind' skills therefore served them well; it helped them become successful, especially in this particular case. They were granted the project funds in the end.

CONCLUSION

In this chapter I have described the practices of collaboration of a group of co-workers in a small Moroccan NGO. I have depicted them as wayfarers of a certain kind, namely as a community of practice, in so far as they mutually engaged in joint activities, continually refining their work, negotiating its meaning and reifying it along the way. Their coming together in the framework of team meetings, and their exchange and development of ideas, made possible an education of attention which was also to a significant extent directed at a practice of perspective taking, which I see as a continual training in the exercise of their 'theory of mind'. Examining the goings-on during one week in particular, I discern a dynamic process of social learning, which was facilitated through instances of joint attention in the task of preparing for an important meeting with a potential funding agency. This involved frequent discussions, brainstorming meetings and joint text-drafting. The preparation work culminated in an intense rehearsal situation (described at the outset of this chapter), which took place one day before the decisive meeting, and during which the team members engaged with the reified product of their joint effort, a slideshow presentation. In this situation of intense collaboration, team members pooled their knowledge and directed it at perfecting the slideshow.

This entailed both a high level of shared intentionality, based on joint attention and strong common ground, and that the collaborators each brought their own vision and expertise into the communicative encounter, thereby enriching the overall picture with the help of distributed cognition. Ongoing social reasoning and joint as-if cognizing during this situation drew on their 'theory of mind' skills and helped them analyse the expectations of their future audience and perfect their product accordingly.

At least two kinds of conclusions can be drawn from this exercise in reflection. One has to do with human life, seen as a bio-social process of becoming. The other concerns sustained participant observation as a valuable resource for the creation of knowledge about this process.

Following the logic to which most authors in this volume adhere, any group of people is 'bio-social' by definition. 'My' team is special, though, in that they may be considered particularly close bio-socially, with the scene I discussed at length representing a peak moment. This is so because by virtue of their long years of working together they have virtually grown and are still growing together – both in the sense of moving towards each other and growing in parallel. Through continuous processes of social learning their brains have grown in tune with each other. Thoughts and feelings, to speak with Bloch (2007), have been going in and out of each other's bodies, rendering their boundaries indeterminate. Engaging in similar body movements, often imitating each other, may also have contributed to further somatic convergence; especially so if one is to follow research on mirror neurons in humans (Rizzolati and Sinigaglia 2006, Iacoboni 2008): their mutual observation behaviour may have contributed to similar neuron activity in their brains, not to speak of other somatic similarities resulting from the consumption of the same or similar food. So whether we call this group a community of practice, wayfarers, an ensemble of bio-social relations (Palsson, Chapter 2, this volume), or co-inhabitants of an ecological niche, it seems to me that to be in the world in the sustained company of others may result in the continuous alteration of our whole bodies.

This brings me to my point about ethnographic fieldwork and engaging in participant observation. Participant observation, by definition, is realized in a community of people. One has to spend a lengthy period of time with a group of people in order to listen to what is said and to make sense of what is being done and, ultimately, also to understand that which is not said, or in short, to grasp how they think (Bloch 1998). By virtue of my sustained presence with the team, of my engaging with them, working and conversing with them, I had begun to understand how they think. Through ethnography I had, in other words, gradually developed a fair amount of common ground with them and had by virtue of continual co-presence also invested with them in considerable joint attention. Also, I had invested in personal relationships, created ties, and formed trust. This continued co-presence had a bio-social effect on me which at the same time offered me the opportunity to gain insights I would not otherwise have received. Immersion with them deepened my understanding of their world. Such bio-social immersion in ethnographic fieldwork is at least one prerequisite for the creation of anthropological knowledge, which is all the more valuable for the advancement of our understanding of what it means to be human if it takes into account the continual interplay of social relations and embodiment (Bloch 2005: 1–19, Ingold 2011: 229–43).

To close, I would like to return to the idea of 'getting one's sea-legs' (cf. Palsson 1994) in ethnography, and to the rehearsal situation. The situation in the converted kitchen definitely struck me as extraordinarily memorable and I suspect it was here that I was put on my sea-legs. Though it was certainly not the only time the team-members experienced a similar condition, it was the only time I witnessed it as intensely during my time with them. The situation in this very small room forced me to belong and to be an integral part of the team; different behaviour would not have been tolerated, taking notes or leaving the room was out of the question. More than at other times, I tried to immerse myself in their way of reasoning, to help analyse the few sentences that were visible on the screen in the light of the social situation ahead, possibly to offer insights from my own kind of folk psychology about how people tend to react to whatever they are presented with. At the same time I was aware of the fact that I had to concentrate on a meta-level as well, to take it all in so that I could later write down what I had experienced in my fieldnotes. In other words, I was extremely alert, even to the point of physical exhaustion.

But the effort was worth it. In this moment I felt, with all my senses, that we were indeed a unit, and that, considering that this situation was not so extraordinary for the others, they actually functioned regularly in this way. This had occurred to me earlier of course – not least because the team leader encouraged the others to work as *un cerveau ensemble*, as a collective brain – but before that day I had pushed the idea aside, thinking that it was mere propaganda or part of a strategy that the leader used in order to motivate everyone else.

I thought I should rather look at them as competing individuals. After all, many of them had served as individual authors of books and I witnessed frequent tension and debates among them. But what I ignored, when I thought that way, was precisely what made practice among this group so special: namely, that they continually sought to integrate what they knew individually so as to construct new knowledge; not only in this particular situation, but in their regular meetings, and for the past ten years at that. On that day I realized that there was no contradiction: there was agency among them, there was friction and contradiction, but this particular group, voluntarily and very effectively, had learned to function as a whole, for individual as well as group benefit.

ACKNOWLEDGEMENTS

Portions of this chapter were presented at the EASA meeting in Maynooth, August 2010, at the ISA research forum in Vienna, March 2011, and at UCLA-CLIC, April 2011. Also, earlier drafts were read by a number of people. I benefited enormously from comments and am extremely grateful. I wish to thank the following in particular: Maurice Bloch, Alessandro Duranti, Andre Gingrich, Tim Ingold, Tami Kremer-Sadlik, Elinor Ochs, Gisli Palsson, Eugenia Ramirez-Goicoechea, Brigitte Vettori and Wolfgang Zeglovits. Last but not least, my greatest thanks go to all members of the team. I sincerely hope that I did not betray your trust.

GAETANO MANGIAMELI

8

The habits of water: marginality and the sacralization of non-humans in North-Eastern Ghana

INTRODUCTION: SELF-WRITING TEXTS

At the end of the nineteenth century, the American writer and naturalist Henry David Thoreau (1906) made some remarks that would resonate with the thinking of both Bronislaw Malinowski and Clifford Geertz. This can be seen by comparing the way Malinowski and Geertz used the term 'culture' with the way the word 'nature' figures in the works of Thoreau. Like Malinowski, indeed, Thoreau aimed at 'grasping the native point of view'. His 'native', however, did not coincide with the Trobriand islanders of Malinowski or with any other indigenous people, but with *nature*. He wanted to become so intimately connected to nature that his observations would eventually become introspective (Worster 1994). It is as if he foresaw the experience of a mutual and emotional relationship between observers and observed to which anthropologists aspired during the second half of the twentieth century. Furthermore, Thoreau was worried about how to read 'nature's poem'. Nearly one century before Geertz (1973) had proposed his celebrated definition of culture as text, Thoreau (1906) lamented the fact that too many pages of nature's poem were missing, unreadable or damaged. What Geertz would have seen as a key feature of culture, for Thoreau was evidence of the anthropogenic decay of nature.

Thoreau was looking for a pristine, uncontaminated landscape seen as a text written once and for all, but he could not find it. His quest for a complete version of nature's poem was the ideal counterpart of

Biosocial Becomings: Integrating Social and Biological Anthropology, eds T. Ingold and G. Palsson. Published by Cambridge University Press.

the notion of culture as a static, internally homogeneous and balanced whole. Thoreau was thinking of nature's poem in terms of being. Had he thought in terms of becoming, his frustration would have been groundless. I suggest that the metaphor of the landscape as text, which has probably been overused in recent years, might profitably be relaunched within current environmental anthropology in the framework of complexity. In order to do this it is necessary, first of all, to identify the environment, rather than the landscape, as the metaphorical referent of the 'text'. Secondly, it would be helpful to focus attention not only on the process of reading, but also on that of writing. Thirdly, we should abandon the common notion of text as a written verbal trace and think of it, instead, as the result of elaboration and diffusion by several different users of an open-source software. That is, we could examine the text by reflecting upon its never-ending rewriting by multiple authors. This is what I mean with the notion of 'habits of water', which has nothing to do either with water as a resource (as it might be taken in an environmentally deterministic approach) or with water as a symbol of life (as it might be understood in a more interpretive one).

The reflections I offer in this chapter are based on fieldwork I carried out in the Kasena chiefdom of Paga, in the Upper East Region of Ghana, between 2003 and 2009. The Kasena are an acephalous, patrilineal and virilocal people settled in North-Eastern Ghana and in Southern Burkina Faso. Mostly farmers, they live in dispersed settlements (*sam*; sing. *songo*), which are grouped together in exogamous sections called *nawuura* (sing. *nawuuri*). Each *nawuuri* has its ritual centre in an 'elder compound', the *sokwio* ('house-father'), namely the *songo* where the elder of the section (*nakwe tu*) lives and in front of which the ancestors' shrine (*naabari*) is located. In short, each Kasena chiefdom comprises the ensemble of the *nawuura* that are loyal to the same paramount chief, the *pe* (literally meaning 'to collect'). Besides the cult of the ancestors, the most prominent ritual elements of local culture are the *kwara* (a horn in which the power of the chief resides; Dittmer 1961) and the *tangwana* (literally meaning 'the skin of the land'; sing. *tangwam*). The latter, which are sacred places consisting in trees, ponds and rocks, belong to a category usually called 'earth shrines' in the anthropological literature on West Africa.

Since these 'sacred groves' have been recently reinterpreted, in the context of the debate about the ecological crisis, as traditional tools for conserving biodiversity, it is worth considering whether this purpose is relevant to the Kasena. This would imply, first, the presence of a cultural conviction according to which human beings are obliged to protect the environment (owing to the compulsory character of sacredness) and, secondly, their ability to do so. Unfortunately, there is nothing like that in the local discourse on sacred places: human beings cannot decide what deserves to be protected. On the contrary, sacred non-human entities *emerge and decline*, wax and wane, in spite of human attempts, ambitions and projects, and call for human attention. The local discourse sees sacred non-human entities as imposing their own presence in the world and ambition to permanence. In order to explain the sacred in this context, the emphasis must be put more on sacralization than on sacredness, more on process than on essences. It is therefore necessary to assume that things must be described in terms of what they *do*, or in terms of how they are made and re-made, and furthermore, in terms of how they make themselves. The Kasena version of the 'sacralization of nature' is totally embedded in the unpredictable and multifactorial process through which the world takes shape, day after day, and in which the human is a relevant component but not the director, since there is no director. No-one is in control of the whole, least of all humans. As a process, sacralization may be summed up in the phrase 'the habits of water', which epitomizes the world in becoming as a collective creation.

Water has no habits, one could say. It just flows; it comes and then quickly goes away, unless its course is hindered by something. The concept of 'habit' seems to be connected to human beings, as it commonly calls to mind the complex ensemble of cultural and individual variations that shape human behaviour. With some effort, a 'modern western thinker' might be willing to attribute habits to non-human animals, but only so as to make indirect reference to instinct, a concept that allows one to recognize forms of order and repetition in the behaviour of non-human animals. This, however, still requires us to uphold a fundamental difference: humans *construct* culture and habits, non-humans *depend on* nature and instinct, despite what much current scholarship on animal culture might suggest. Certainly, in order to extend the notion of 'habit' to abiotic phenomena such as water, wind or fire, it is necessary to expand the metaphoric endeavour of the writer and to ask for the interpretive indulgence of the reader. With the word 'habits' we usually label something foreseeable, something that has already happened and that is supposed to happen again. As anthropologists, however, we are also interested in the mechanisms that make this 'something' happen: if habits, in general, are inserted in a flux that connects them to what they give rise to, on the one hand,

and to what generates them, on the other, then the habits of water emerge from a similar spiral.

After all, water adapts itself to the surface of the land, but at the same time modifies the surface itself. Like the *habitus*, which is a system of structured and structuring dispositions (Bourdieu 1972), the flow of water is moulded by those external conditions that it contributes to moulding at the same time. By digging its own way into the soil, rainwater discriminates between spaces that are suitable for building houses or farming and spaces that are not. If rain itself is totally beyond control, its path on the surface of the land is not, as it can be handled, whether purposefully or by chance.

Besides water, together with it and against it, both humans and non-humans work on the *skin of the land*; they can care for it and modify it. Landscapes are meaningful for human purposes when water is not *everywhere*, but rather *somewhere*, as there is no meaning without difference. Indeed, discontinuities on the surface of the land are precisely what humans need in order to turn the landscape into a readable text. Yet unlike the way books are commonly conceived, this kind of text is not the work of any isolated author. Landscapes are continually co-produced by a plethora of authors, no one of which, as an individual, is definitively responsible for what results from the writing. Such an activity is always ongoing and produces arrangements whose nature is necessarily temporary. So, like the *habitus*, the flow of water is 'orchestrated without being the product of the organizing action of a conductor', as Bourdieu (1972) puts it. The world writes itself, as in Escher's famous lithograph, *Drawing Hands*.

THE CENTRALITY OF MARGINALITY IN BIOSOCIAL BECOMINGS

The definition of anthropology as the study of biosocial becomings implies a double operation of reframing with regard to both the subject we observe or interact with in order to comprehend, and the horizon of comprehension itself. First, the anthropology of biosocial becomings abandons the focus on single entities and their essential features to move to the level of the ensembles of biosocial relations they form together. Secondly, it moves from actuality to potentiality. In other words, anthropologists can open up not only to *what happens*, but also to *what else might have happened*, given that a good way to catch meanings consists in looking at the gap between actual events and potential alternatives.

Defined in this way, anthropological inquiry would focus on the systems of interaction and exchange through which events occur, in order to detect the possibilities opened by a plethora of movements that are not coordinated by any higher 'mind' or centre of power. Besides the strict alternative between attributing agency to non-humans and the opposite, more orthodox view according to which only humans have agency, there is a third way: agency and creativeness, in the sense put forward by Whitehead (1926), belong to ensembles of interacting entities or eventually to complex adaptive systems as defined by Lansing (2003).

Narratives of equilibrium or natural balance fail in understanding life, both at the level of the individual human body and at the level of communities of humans and non-humans. Although disequilibrium and uncertainty are key features of life processes, they are still commonly perceived as imperfections or malfunctions that need to be fixed. The anthropology of life, on the contrary, is an attempt to understand human beings in the contexts created by their interactions with other entities, and more precisely in terms of what they do rather than what they are, which, in turn, makes the use of the verb 'to be' somewhat slippery. Our experience in the world provides several examples of *whirls of organisms*: spirals of life in which humans and non-humans are entwined. In this framework, the basic subject of anthropologists' interpretive efforts is constituted by multiple levels of interaction that do not necessarily work in harmony. In other words, it is constituted by multifactorial processes whose unpredictable outcomes are the temporary ground upon which we try to grasp meaning through observation, while underlining the importance of marginality. Here I am referring not to marginality in the classical anthropological sense, namely as anti-structural or liminal (Turner 1974), but rather to the meaning it has assumed in economics: marginal as an increment, an increase or a decrease, something that we all can make every day, producing differences in an almost imperceptible way.

If the move from actuality to potentiality requires us to include not only 'what happens' but also 'what might happen', then the definition of anthropology as the study of biosocial becomings makes it necessary to consider these becomings in the wider context of open systems of interspecific interaction. In other words, an understanding of *human* becomings calls for an ethnography of *non-human* becomings, not just to add a 'bio' (the non-human) to the 'social' or 'socio-cultural' (the human) – an intellectual operation that would confirm and reinforce the biases of modernity – but more precisely to recognize the value of contingent forms. If we work and think in an essentialist frame, then contingency is simply the lack of permanence in the context of 'the-world-as-it-is' (or rather, 'the-world-as-we-think-it-is'). If however we focus on becomings, contingency opens the way to the unfolding of 'the-world-as-it-might-be'. The role of anthropology itself becomes political if anthropological writings cast light on and promote the free interaction of elements through which new collective enterprises can arise. This is just another way to make creativity central instead of treating it as the extraordinary capability of a single genius. Marginal acts are, again, the necessary condition for any form to emerge.

While anthropologists attempt to reconstruct their frameworks and discourses to pose new intellectual challenges, it is worth describing contexts in which people are already involved in conceptual experiments around biosocial becomings. During my fieldwork in the chiefdom of Paga, I found that I could use the notion of the sacred as an interpretive tool to think about what might be called ensembles of biosocial relations (Palsson, Chapter 2, this volume). Local people describe sacred entities by what they do, not by what they are. This implies, in anthropological terms, that knowledge of these entities concerns how objects, taken in the broadest sense, are made, which then raises questions about 'who' makes them.

In order to show what the anthropology of biosocial becomings might look like, I will start from the antithesis of 'becoming', namely 'being'. The 'to be versus to become' question is crucial in so far as it entails the notion of 'sacred', which, in turn, is the key to understanding the vision of the world inscribed in Kasena religious institutions. With regard to the so-called Earth shrines among the Kasena, an interpretation of the sacralization of nature in terms of becoming is effective whereas an enquiry into being is not. This contrast can be addressed by introducing three enigmatic aspects of relations between humans and non-humans. First, according to local conceptions, the baobab is the clearest example of a sacred tree protected by a taboo. As is commonly said, 'baobabs are God's trees. We don't plant them, we don't cut them'. At the same time, Kasena claim that baobabs, as any other trees, can be freely cut down when they are 'young'. Second, when the Kasena want to offer a sacrifice to one of their sacred ponds, they perform it elsewhere, in nearby groves, and they explain this choice by saying that the pond and the grove are the same place. Third, the sacred crocodiles of Paga are widely known for their harmless attitude towards human beings. In recent years, however, it has begun to be widely held, among neighbouring peoples, that 'the harmless crocodiles of Paga are getting wild'.

A framework for the Kasena understanding of their lifeworld in terms of becoming is provided by their concept of creation. According to local cosmogony, everything in the world traces its origins back to the *dwi*, the flesh of a 'cosmic calabash' (Liberski-Bagnoud 2002: 102), through a process of progressive differentiation. In other words, creation consists in an ongoing process of autopoiesis, and not in the application of God's intelligent design: the world has created itself. As a result, all entities, as components of a complex adaptive system, have a meaning immanent in the unfolding of their properties through interactions in time, and are interconnected in so far as they spring from the same process. It is on this basis, first of all, that analogies between different entities can be drawn in order to build a biosocial understanding of life.

In fact, two opposite cosmogonies currently coexist in Kasenaland, which have not been in conflict so far probably because no one has put the issue on the agenda. They are epitomised by two irreconcilable descriptions of the supreme god, *We*, the Sky. The first, which seems to be in contrast to the concept of *dwi* and might have been recently borrowed from Christianity, sees the high god as the Creator. *We* is a *deus otiosus* dwelling in the sky, apparently not interested in what happens on the Earth. I will call it 'God-as-the-Creator'. The second one, like the first, is a *deus otiosus*, but this god is not the Creator: it is associated with the sky not because that is the place it inhabits, whence it has created the world, but because it has ascended to the sky from Earth, where it formerly dwelled. This description of God is consistent with the cosmogony according to which the world has created itself starting from the cosmic calabash. Accordingly, God is just the most prominent element produced by this process of self-creation of the world. It has become what it appears to be. I will call it 'God-as-the-Emergent'. In neither version is the high god involved in any process of creation, but for opposite reasons: God-as-the-Creator does not create as it has already created everything once and for all; God-as-the-Emergent does not create as it is rather part of an ongoing process of creation. Recalling Thoreau to mind, his frustration would have been appeased if he had thought of God-as-the-Emergent, one of a plethora of authors and a formerly engaged reader of nature's poem, rather than of God-as-the-Creator, the sole author and detached observer of the world.

The modern sciences are characterized by what Ingold has defined the 'double disengagement of the observer from the world', consisting in 'a division between humanity and nature', on the one hand, and in 'a division, within humanity, between "native" or "indigenous" people ... and enlightened Westerners', on the other. Thus, according to Ingold, 'the sovereign perspective of abstract reason is a product of two dichotomies: between humanity and nature, and between modernity and tradition' (Ingold 2000: 15). Furthermore, the separation of 'humanity' and 'nature' into two different houses is at the core of Latour's definition of the 'Constitution of Modernity'. The first house is overcrowded with 'the totality of speaking humans, who find themselves with no power at all save that of being ignorant in common, or of agreeing by convention to create fictions devoid of any external reality', while the second is the house of non-humans, 'real objects that have the property of defining what exists but that lack the gift of speech' (Latour 2004: 14). The illusion of modernity is challenged, in the case of sacred things among the Kasena, in two respects. First, non-human entities which, in the context of modernity, are supposed to be speechless entities that lack the ability to define social reality, appear on the contrary to be the source of moral commentaries upon human life. Secondly, as the 'double disengagement' of the 'modern observer' is clearly revealed to be a chimera, there is nothing that human beings can observe without being entangled with it.

Among the Kasena, all sacred non-human entities have something to say. They are described as subjects endowed with rights and the power to act, and they have their own views concerning everyday human activities and choices. Together with their human counterparts, they form what Latour would call an assembly of humans and non-humans. As the composition of this assembly – the list of sacred places – is always provisional, its meaning is elusive. New gods are discovered, others are forgotten or destroyed for a number of reasons, the most prominent of which is the possibility of a compound's being abandoned (and then, as time goes by, the gods located in the surroundings are abandoned as well). The activity of the Batesonian ecological mind (Bateson 1972, 1979), which has been conveniently redefined by Ingold (2000: 16) as an ecology of life, is immanent in the relations between its components, which reveal themselves through local interactions, and consist precisely in recognizing, confirming or removing the non-human entities that cooperate with humans in building an understanding of life. In a context where no wall separates natural objects from socio-cultural subjects, it is neither possible to conceive of a bounded realm of nature, nor would it appear relevant to make any generalization about natural species on a genetic basis.

In short, non-human entities are meaningful in so far as they are engaged in relations with humans. With reference to Ingold's (2000: 209–218) well-known distinction between globes and spheres, sacred non-human entities are perceived through a local perspective that is focused on them as organisms in specific locations and not as occurrences of the species they belong to. This is true with regard to sacred trees, which are always very close to human settlements, as well as to crocodiles, which are harmless only when a specific kind of engagement with humans occurs. In other words, enlarging the production of meanings beyond the limits of humanity does not imply either that everything is meaningful or that all non-humans are subjects. A kind of difference is always necessary. Otherwise, there would be no meaning.

Life as a collective creation based on marginal acts is epitomised by the *puru*, which is a shrine located in front of the entrance to Kasena compounds. Its size is a measure of the compound's prestige: the bigger it is the greater the prestige. Despite the relevance of the *puru*'s size, upon which the reputation of the compound's inhabitants depends, the latter do not simply make the shrine bigger through some technique of permanent construction. In fact the *puru* is built every day, slowly, through marginal acts. Basically, the *puru* is a heap made of waste: it is the place where every morning, women pile up dust and rubbish after they finish sweeping. Its aspiration to stability and uprightness is made possible by the grass growing on its surface, yet it is hindered by the erosive power of wind. Rain, on the other hand, behaves with some ambivalence: it can make the heap more compact but also make it slide down. Furthermore, every time a newborn comes to life in the compound, a pot filled with the baby's placenta is buried in the *puru*. In the same way as compounds can grow, decline, and sometimes disappear, the heap can grow or decline, depending on the intersection of atmospheric conditions and demographic trends in the compound. Finally, there is a third possibility of human intervention in the *puru*, but it is only occasional. The head of the compound, no more than three times in his life, can ask his people to restore the *puru*. This limit seems to underline the fact that humans are not allowed to stack the deck. The decay of a compound cannot be denied by repeatedly restoring the *puru*, or in other words, by reinforcing the conscious purpose, for which there is no place in a complex adaptive system.

The *puru* is built by life processes and possibly erased at death. Intended as a monument to social life in a parcel of land, it is – like life itself – involved in a continual becoming without a precise direction. The *puru* embodies the centrality of life-in-the-making in a West

African philosophy of becoming, in which the only way to understand what human and non-human entities (including artefacts and 'natural' phenomena) *are* consists in describing *the things they do*, and how they are made and re-made.

TREES ARE POTENTIALLY NATURAL

In Kasena-land, the Earth shrines, *tangwana*, are literally 'the skin of the land'. Local commentaries see each of them as 'something that comes out to reveal a secret'. The secret they reveal is actually nothing but their own presence in the world and, as such, is related to the environmental conditions of the savannah, which can be easily misunderstood, as Fairhead and Leach (1996) have shown.

Despite the widespread conviction according to which the sacralization of trees and ponds is a way to celebrate life in keeping with a long-standing fear of drought, I would suggest that the problem is of a different nature. The association between the savannah and drought is based on a static, photographic imagery that allows one to see only one side of the environmental issue. The other side can be seen if one looks at the savannah in its changing appearance throughout the seasons. The main ecological concern for savannah people is the ambivalence of water, since it poses a double threat to life: water is extremely scarce during the dry season and abundant during the rainy season, when it can destroy what humans and non-humans have built. From this point of view, it seems clear that interpretations of sacralization based on so-called 'relic theory' (see Sheridan 2008: 10–14, for a critique) are unreliable. Sacralization has rather to do with a basic problem for both humans and non-humans: how to choose a safe place to stay, which has to be close enough to water and at the same time safe from water. It is about the relations between the habits of water and the skin of the land.

What makes a tree sacred? Since the Kasena, when they speak in English, state that their sacred places are *natural*, it is worth describing how nature and culture are locally understood. In the Kasena worldview there is no room either for the nature/culture dichotomy or for the division of intellectual labour that depends on it. *Kasem*, the local language, has no words for nature or culture. Nevertheless, owing to contact with the British, the Kasena have imported the concepts of 'natural' and 'cultural' by reinterpreting them in the framework of their philosophy of becoming. The results of this reinterpretation of essentialist concepts in terms of relational ones are interesting, if not ironic. The word *chullu*, which is used to translate 'culture', covers all objects, living beings and actions that are involved in ritual prescriptions or prohibitions, the foundation of which is always credited to non-humans. In other words, to a large extent, what is understood as 'culture' is constructed by non-humans. The 'natural', by contrast, despite its Latin etymology, is understood as a condition that has to be attained. Even though they have no notion of a bounded realm of nature, the Kasena offer two different understandings of the adjective 'natural'.

The first is the concept of *We teni* ('God's trees'; sing. *We teo*). According to this principle, only non-cultivated species of trees are natural, as they grow of God's will and not because of human commitment. Here, the supreme god represents the a posteriori acknowledgement of the unpredictable dynamics of complexity. In spite of what such a definition of the natural would suggest, an explanation of sacred trees based on a description of the species they belong to is destined to fail. In recent years, sacred groves have been understood by environmental activists as indigenous ways of conserving biodiversity. From this point of view, sacralization could be interpreted as the application of a cultural project to the natural world. Humans would be detached observers of a non-human realm regarded as material, objective and biologically determined. If this were so, a discourse of being would be appropriate, since local knowledge of species would underlie the selection of sacred trees.

But this is not in fact the case for the Kasena. Here we find the clear relevance of Ingold's point that the very concept of biodiversity entails a global perspective (2000: 217). Local people know more than fifty species of trees in their territory, nearly all of which are considered *potentially sacred* – that is, *potentially natural* – in so far as they are not cultivated. Nevertheless, if we map all the sacred places currently known, we discover that the trees involved, here and now, belong to a very limited number of species. This happens because new gods are discovered every time someone is planning to clear land in order to build a house or to farm. The Kasena recognize as gods only grown-up trees standing close to their settlements. If we observe a sapling – if we look at its biological features as a being – then nothing can help us predict whether it will become a sacred tree or not. Later on, if the sapling has not been swept away by water, eaten by animals, destroyed by children while playing or cut by farmers for any reason – if, in other words, it has gone through a long chain of marginal events without dying – then it might have the chance (but never the certainty) of being recognized as sacred, namely as natural. In short, sacralization occurs when a tree has already shown itself capable of growing in a hostile environment. This is what Kasena denote as 'natural' (and as 'sacred', at the same time): unplanned and unpredictable. Trees become natural in so far as they have solved the same problem with which humans are concerned: building something that aims to permanence. In other words, they have, first and foremost, managed to build themselves. Behind the sacralization of trees there is neither a merely biological understanding of species, in essentialist terms, nor any kind of super-organic cultural project superimposed upon material reality, but a process that involves both form and substance.

The second definition of 'natural' is based on the concept of *naboro*, which is related to the ancestors, as *naabaaro* means 'ancestor' and *naabaari* means 'ancestor shrine'. *Naboro* is used to label all entities that are assumed to have existed since ancestral times. In this case 'natural' seems to be understood as 'native'. From this point of view, the cult of sacred places might be interpreted as related to the conservation of pristine landscapes, yet sacred ponds are far from being ancestral or untouchable. First of all, when a sacred pond becomes dry, it sometimes calls for human support through divination. In such circumstances, humans dig in order to help the pond restore its minimum water level and meet the needs of the sacred crocodiles that live there. Through this ritual procedure, humans contribute to the daily work of all the trees surrounding the pond in reinforcing the embankments. Furthermore, it is well known that during their colonial rule, the British launched a plan aimed at increasing the water supply in the Northern Territories of the Gold Coast by building dams. This programme led to the creation of huge ponds that are now worshipped as *tangwana*. The innovation brought by the British was simply acknowledged as part of human history which, locally, is not interpreted as separated from 'nature'. Local people openly share information about how they worked to modify the landscape and how such modifications resulted in the discovery of new gods. Nevertheless, they claim that these ponds are not man-made: 'We just dug. Water was already there'. Moreover, since their sacred crocodiles moved from smaller ponds to the new ones, and since trees started to grow on the embankments, the whole complex of 'artificial' ponds could hardly be considered anthropogenic.

It is worth noting that the two aforementioned definitions of natural do not necessarily overlap. In this respect, the case of the neem tree is highly revealing. On the one hand, as the neem tree was deliberately introduced to Ghana during British colonial rule, and consequently cannot be considered *naboro*, it is not natural. In fact, in the Kasem language, the neem tree is called *fela tio*, meaning 'white man's tree'. On the other hand, since neems have been proliferating on their own, they are in the process of *becoming native*, as I was told. All the neem trees surrounding ponds (but these alone) are currently recognized as sacred trees, in so far as they help the *tangwana* control the flow of water. In short, for a tree there is no necessary relation between being a *We teo* and being a *tangwam*, yet even trees that are not *naboro* can attain the condition of *tangwam*. Although refined in terms of medical and culinary applications, Kasena ethnobotany does not help us in understanding the notions of 'sacred' and 'natural', for the taxonomy informs us about beings, whereas sacralization is related to unpredictable becomings.

SUCCESSES AND FAILURES

The second and the third enigmas to be clarified are strictly interconnected: how can a grove and a pond be considered 'the same place' – allowing sacrifices performed near the former to be destined for the latter – and how can crocodiles become wild?

According to local belief, crocodiles *are* human beings on the basis of a local history of mutual engagement. It is the latter (the process of becoming) that explains the former (the being), and not vice versa. The myth of foundation of Paga tells how the ancestor made a deal of mutual friendship with a crocodile. On this basis, humans were forbidden to kill crocodiles, to eat their meat and even to harm them in any way. The paramount chief of Paga, as the founder's patrilineal descendant, guarantees the observance of the mythical pact between humans and crocodiles. Nowadays, each crocodile in Paga corresponds to one human being, and vice versa. To be more exact, according to the Kasena concept of the person, human beings have many spiritual components, but only one, the individual mortal soul, *juru*, literally meaning 'mirror', is shared by humans and crocodiles: this is what local people mean when they say that crocodiles are their souls.

When asked to compare older and younger humans, older and younger compounds, humans and trees, or trees and animals, the Kasena would usually say that 'they are different, but they are also the same'. As Descola puts it, West African analogism acknowledges a plethora of differences and looks for similarities that are not to be found in the entities themselves, but rather in the relations between them. In this context, as he has argued, sacrifice is the instrument through which continuities between intrinsically different singularities are instituted (Descola 2005: 280–281). Sacred ponds and sacred groves are living examples of two different kinds of relations, between the skin of the land and the habits of water. In the former case, soil surrounds and contains water; in the latter, water surrounds patches of land without invading them. Here and there, trees are committed to building discontinuities on the surface of the land, namely to writing differences on the landscape as text. Humans need both categories of space: one to collect water, the other to live in. Sacred groves are ideal locations for future human settlements and, in some cases, they have grown on the ruins of abandoned settlements. If humans clear land to build their compounds, non-humans can afterwards reclaim this space. Nevertheless, no location (neither for humans nor for non-humans) lasts forever, as it needs the long-term repetition of marginal acts of maintenance. Trees must keep on living in order to circumscribe the pond, and humans must keep on inhabiting the house in order to keep it in repair. By choosing groves for the performance of sacrifices destined for ponds, humans connect their potential or past locations to the places where crocodiles, their non-human counterparts, currently dwell. Furthermore, at the end of the sacrifice, they throw part of the meat into the water to feed the god – and the crocodiles, because they *must* share the meat with their sacred mirrors.

The process whereby crocodiles have become harmless – a process that we must understand if we are to know how they can become wild – began precisely from this specific aspect of the Kasena procedure of sacrifice. At the beginning of the twentieth century, with the end of the slave raids which had been a major threat to peace for about three centuries, the Kasena established new relations of exchange with Muslim cattle traders from the north, who needed safe places to rest and water for their cattle while travelling south. In return for their services, the Kasena received heads of cattle that were to be sacrificed to the ponds. As a result, crocodiles were offered meat acquired through a new form of regional economy. Even though this procedure had already been long established at the time, the frequency of transactions with cattle traders, and consequently the abundance of offerings to the gods, was unusual. It did not take long for people to realize that crocodiles were becoming accustomed to receiving meat from humans. Step by step, after this fortuitous discovery, humans and crocodiles developed new skills of mutual interaction. By the end of the last century, specifically trained crocodiles had already become a kind of local tourist attraction for strangers. Today, in the context of a community-based eco-tourism project, tourists (and sometimes anthropologists) from all parts of the world meet the crocodiles of Paga, take pictures of them, hold their tails and sit on them. At the end of the show, and so long as they behaved properly, the crocodiles are fed fowls. In other words, the performance is part of the ongoing training of the crocodiles. More importantly, when they talk about this subject Kasena usually recall and underline the role of a number of different protagonists in this story (slave raiders, cattle traders, elders, crocodiles, *tangwana*, NGO staff, British and French soldiers and so on), none of which, taken alone, has full responsibility for any specific cultural project related to crocodiles.

If the harmless attitude of crocodiles is a side effect of a moral economy of sharing between humans and their gods, the same can be said of the inverse process, through which crocodiles are becoming wild. Even though crocodiles usually behave as they are expected to, accidents are not infrequent. In recent years, when crocodiles happened to attack humans, this was locally interpreted as a direct consequence of humans' misbehaviour. Rumours of embezzlement committed by tourism staff are becoming increasingly common. According to informants, some staff members kept the money tourists had paid to feed the crocodiles at the end of the performance. In these cases, humans are not sharing with crocodiles the income resulting from the display of their reciprocal engagement. If mutual responsiveness is a key factor in co-evolution, then it is seen at play both when humans respect their pact of sharing things with the crocodiles (with positive results) and when they fail to do so (with tragic outcomes).

CONCLUSIONS

In one of his metalogues, Bateson (1972) discusses with his daughter the game of croquet described by Lewis Carroll in *Alice's Adventures in Wonderland* and underlines the 'muddle' stemming from the use of flamingos as mallets and of hedgehogs as balls: as the trajectory of the ball, to a large extent, depends on flamingos and hedgehogs, rather than only on the players' skills, the result is unpredictable (Bateson 1972: 40–41). On the other hand, Lévi-Strauss (1962) defines ritual as 'a favoured instance of a game . . . the only one which results in a particular type of equilibrium between the two sides' (1962: 30). If a ritual, from this point of view, might be interpreted as a session of a game whose uncertainty has been domesticated, then it is worth trying to compare Bateson's game and Lévi-Strauss' ritual with the Kasena's game of chance. Here, instead of flamingos and hedgehogs we have crocodiles, and in place of the trajectory of the ball it is the safety of players that is subject to uncertainty. Show, game or ritual? Perhaps, all three together.

Certainly, as visitors are directly involved, this performance is more than a show or an exhibition. More importantly, the performance is surrounded by an aura of uncertainty that constitutes the main reason for its interest. More precisely, the game of chance is characterized by at least three degrees of uncertainty. First, will the crocodiles come out of the water to meet their visitors? Secondly, will they allow humans to come near to them? Thirdly, will they wait for the exhibition to end and then go back to the water without harming any human? Whatever crocodiles do – even if they stayed in the water without meeting the visitors – would be an answer. As soon as a context of communication is established, it is impossible not to communicate.

Lévi-Strauss's definition of ritual is useful to describe what happens when crocodiles behave as expected: in this case, humans receive a positive answer. The performance itself is a kind of acted statement: 'As you have personally experienced, crocodiles really are harmless'. In other words, no matter how bizarre local beliefs may sound, the performance shows that they are based on empirical evidence. Conversely, it should also be realized that the credibility of this statement, which is at the core of the Kasena identity and ritual system, is at stake – or better, under threat – every time a tourist meets a crocodile of Paga.

Finally, a series of oppositions can be drawn between the beginning of this process and its recent evolution. The performances were first staged as informal practices that anyone could perform in any pond of Paga, on the basis of a personal negotiation with a visitor. Today, after a reorganization coordinated by an American NGO, the performance is exclusively managed by official tourism staff working in two of the huge artificial ponds built in the twentieth century. Summoning crocodiles in any of the several other ponds of Paga is now forbidden, but as the small ponds provide water only during the rainy season and then quickly dry up, over the last decades crocodiles have gradually disappeared from them and moved to where the shows are allowed: the artificial ponds that contain water throughout the year. About a century ago, the moral economy of sharing between humans and crocodiles was connected to the new, open and friendly relations established between local people and strangers after centuries of conflict. Today, discourses about crocodiles becoming wild are nothing but commentaries on how humans are becoming greedy because of the penetration of western money, which is just another way of suggesting that sacred non-human entities always have something to say about human becomings.

To conclude, 'sacred' is used to label some pivotal examples of sociability connecting a number of different agents, whose nature might be human, non-human, biotic or abiotic: trees that overcome all sorts of accidents, thin embankments that keep on containing water, crocodiles that accept fowls instead of biting their human visitors, heaps of waste aspiring to spectacular uprightness, all of them interacting with the everyday marginal steps of humans and other non-humans. In all such cases, the focus of comprehension is not their supposed essence, but rather the description of the actions they undertake and the events they are involved in. Such examples are to be seen as successful outcomes of unpredictable processes, and as such they could not be more clearly contingent than they are. The value of such 'successes' resides in the absence of a single planning mind, on the one hand, and in the possibility of failure, on the other. It is precisely on the thin line separating success and failure, on the line of uncertainty, that sacralization reveals itself to be the celebration of life processes that might be called biosocial becomings or collective creations.

In this context, whatever one thinks about the relation between anthropology and ethnography or about nomothetic or idiographic purposes, anthropological discourse ceases to aim at describing or comprehending the world-as-it-is, and turns to the world-as-it-may-become, revealing itself to be one of the many forces pushing the world in one direction or another.

VITO LATERZA, BOB FORRESTER AND PATIENCE MUSUSA

9

'Bringing wood to life': lines, flows and materials in a Swazi sawmill[50]

As the other chapters in this collection clearly show, the study of biosocial becomings calls for major epistemological shifts in the practice of anthropology. This essay is an attempt to show some of these potentialities, and focuses on the productive processes of a Swazi sawmill, engaging humans, machines, wood, and other materials, as porous overlapping organisms and things intertwined in one and the same environment. The sawmill is located in Enkopolwani, Swaziland, previously an asbestos mining town; it is now being redeveloped by the not-for-profit company Enkopolwani Ministries Swaziland (from now on EMS) as a Pentecostal Christian sustainable business project with the primary mission to provide care for orphans while producing economic wealth from activities such as timber logging and plank production.[51]

Through detailed descriptions of life in the sawmill, the chapter also offers a systematic application of Ingold's phenomenology of lines, flows and materials (Ingold 2006b, 2007a 2007b, 2008, 2009, 2010b, 2011) to an extended ethnographic case study, in dialectical

[50] Vito Laterza is the primary author of the text, which is based on his fieldwork experience in Swaziland between February 2007 and December 2008. The first person in the text refers to him. Bob Forrester is the photographer responsible for the sawmill photos, which performed a crucial role in the development of the text; he also collaborated with fieldwork and analysis. Patience Mususa collaborated in the mapping of the sawmill production process and produced the AutoCAD drawings presented here.

[51] Unless otherwise stated, all names of specific places and persons in Swaziland were anonymized to protect research participants.

Biosocial Becomings: Integrating Social and Biological Anthropology, eds T. Ingold and G. Palsson. Published by Cambridge University Press. Cambridge University Press 2013.

conversation with some of the concerns of Marxist factory ethnography (Beynon 1973, Burawoy 1979, Parry 1999, Mollona 2009). It develops Ingold's critique of agency to move beyond a view of workers, industrial machinery and intermediate and finished products interacting as separate bounded entities. Instead, it favours a perceptual engagement with life that privileges flow, flux and process, grounded in the constant production, transformation and dissolution of materials. A 'material' refers here to anything with intrinsic properties, enabling and constraining specific paths of becoming in the environment (Ingold 2007b: 14). Ideas are – and are made of – materials just as much as is wood or sawdust. No a priori distinction is thus assumed between mind and body, the mental and the material, the social and the biological. Materials, like their becomings, are biosocial.

NETWORKS AND AGENCIES

Ingold argues that anthropologists and other social scientists often tend to capture and contain the movements and ambiguities of social life into neat abstract frameworks that are expected to provide valid representations of real-life situations (Ingold 2007a: 72–75). In the attempt to get at fundamentals, to explain the 'true' dynamics of continuity and change, scholars identify discrete entities, sometimes human subjects and natural objects, or supernatural and even technical agencies, and map relations and causalities among these agencies. The assumption is that no matter how 'messy' life out there might seem, there must be some relatively bounded entities that constitute the core of this 'mess'. What matters are the connections between these entities, or to put it differently, their causal influences. The network is a good representation of this epistemology (Figure 9.1). The power of this imaginary network of points and straight lines is far-reaching.[52] In the midst of the ambiguities and blurred boundaries of experience, it enables the anthropologist to see through the mist of apparent chaos and find regularities, sequences, causes – in short, an underlying order that explains and predicts behaviours and trajectories

[52] This is obviously a simplification. One might object that each point in the network is in itself a conglomerate of other points, embedded in complex networks of causality and interaction. Yet, thinking in points and straight lines is the basic logic in most cases. Even when ambiguities, complexities, fragmentations and overlaps are acknowledged, these are often defined by reference to the boundaries and distinctions they lack, rather than in their own terms – see, for instance, the notion of 'quasi-objects' employed by Latour (1993: 89).

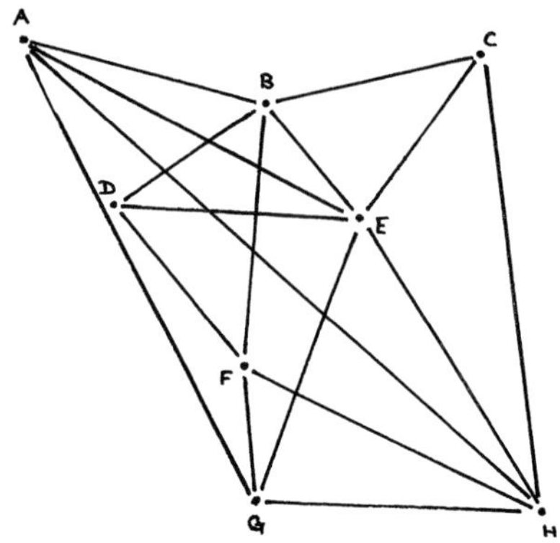

Figure 9.1. Network, from *Lines* (Ingold 2007a: 82); by permission of the author.

of change. Movement in real life, however, is more like a trajectory traced along a continuous wayfaring line. What the analyst tends to do in order to map this is to first break the continuous movement into a sequence of dots, and then to re-connect the dots into segments, that is, with straight lines. This entails a move from the *walk*, an irregular continuous path through the environment, to the *assembly*, an imaginary construction of point-to-point connectors (Ingold 2007a: 72–75).

What is agency then? From the network perspective, it is the attribution of the ability to act to specific entities, human, material, technical or otherwise. In order to explain why things and entities move and fluctuate, one needs an external explanation, some kind of force separate from the movements of life themselves. As Ingold puts it,

> this world can only be brought back to life in the dreams of theorists by conjuring a magical mind-dust that, sprinkled among its constituents, is supposed to set them physically in motion. (2007b: 11)

MARXIST FACTORY ETHNOGRAPHY

At risk of erecting a straw man, a similar logic can be detected in Marxist factory ethnography. The main contribution of factory ethnography to Marxist critiques of capitalism has been to provide detailed descriptions and analyses of socio-technical relations between workers and managers and of production processes on the shopfloor. Their ambition is to advance one of Marx's central concerns: 'the concrete relations of production in the capitalist factory' (Castoriadis 1988: 105). Abstract claims about the forces of capital, exploitation, class conflict, class consciousness and workers' solidarity are here given a face and a place, accompanied by a detailed exploration of the micro-dynamics of capitalism. In practice, however, the Marxist theoretical framework only reproduces itself, with only minor changes from other variants of Marxism not grounded in ethnography. Whatever the theory of capitalist change or class consciousness produced by these ethnographies, the basic categories of workers, managers, capital, class, internal contradictions, conflict, consciousness, and so on, are seldom questioned as valid tools for representing reality.

The Marxist framework only works in so far as one identifies bounded subjects and objects and 'animates' them with some kind of agency. Exploitation, appropriation of surplus value from workers, inherent conflict between workers and managers, are assumed as a priori conditions, and explicated afterwards through ethnographic data. A good summary of this 'concrete' ethnographic approach to Marxism is provided by Beynon's formulation of the concept of the 'frontier of control':

> On the shop floor of many factories the division between the supervisor and the men can be characterized as a 'frontier of control' – management's rights on the one side and those of the workers on the other. It is in this way, in disputes over control at work, that the class struggle has been fought out by the British working class during this century. At the lowest, and most fundamental level, it has involved a conflict over how much work the men do and how much they get paid for it. At its most developed level it has produced an ideological conflict over who runs the factory and why, to a questioning of the essential nature and purpose of production within a capitalist society. (1973: 139)

While these works present important methodological and theoretical differences, they share a common rectilinear logic. In this case, the points are constituted by categories such as workers, managers, capitalists, machines. These are then joined by straight lines that indicate the movements and contradictions of capitalist production. Workers fight managers. Managers oppress workers. The forces of capital try to buy workers' consent, and so on and so forth. The straight lines are crucial to explain movement in real life. Points become alternatively cause or effect. This process can be iterated numerous times, adding to the complexity of the network of point-to-point connectors. Yet the primary logical operation remains the same: draw the dots, join them, find new dots, discard old ones, trace new straight lines, delete other ones. The 'worm's eye field view' (Mollona *et al.* 2009: xi) advocated by these ethnographies is built upon an interactionist conception of entities and forces that falls short of questioning the existence of relatively fixed boundaries among humans, machines and the other 'physical' materials involved in production. Despite these theoretical straitjackets, the descriptions presented by these authors rank among the most vivid and detailed in the anthropological literature (Burawoy 1979: 67–70, Parry 1999: 120–126, Mollona 2009: 23–24). They are essential readings for anyone interested in understanding the real-life processes entailed in industrial production. This is yet another sign that ethnographic writing too, like any other living organism, leaks beyond the boundaries and the intentions imposed by the authors. As such, it remains an indispensable tool for restoring life in the often static writing of social scientific literature.[53]

BIOSOCIAL BECOMINGS AND THE MESHWORK OF LIFE

How does one move beyond agency and separation? A good start is to focus on the wayfaring movements of life in its multiple biosocial becomings. A useful tool to think with is the meshwork (Figure 9.2). If the network of point-to-point connectors is the expansion of the assembly, the meshwork develops out of the wayfaring paths of social life.

From this perspective, human and non-human organisms and things are bundles of wayfaring lines inserted into the wider meshwork. Open-ended and with ever porous boundaries, these knots of materials and activities become visible as concentrations of wayfaring lines in specific areas of the environment. What sustains the permeable boundaries is precisely the fact that biosocial materials – continually meshing together and transforming into yet other materials and substances – are wayfaring through an organism or thing, leading to other knots and activities. This flux, and the tensions that forces and materials generate in the process, sustain the organism in its perceptually concrete existence. The meshwork metaphor enables one to

[53] This critique refers specifically to Marxist factory ethnographies. Numerous anthropological writings influenced by Marx and Marxism do not subscribe to the framework challenged here – see for instance Hornborg's (2001) ecosemiotics or Ingold's (1986) early work on human ecology and sociality.

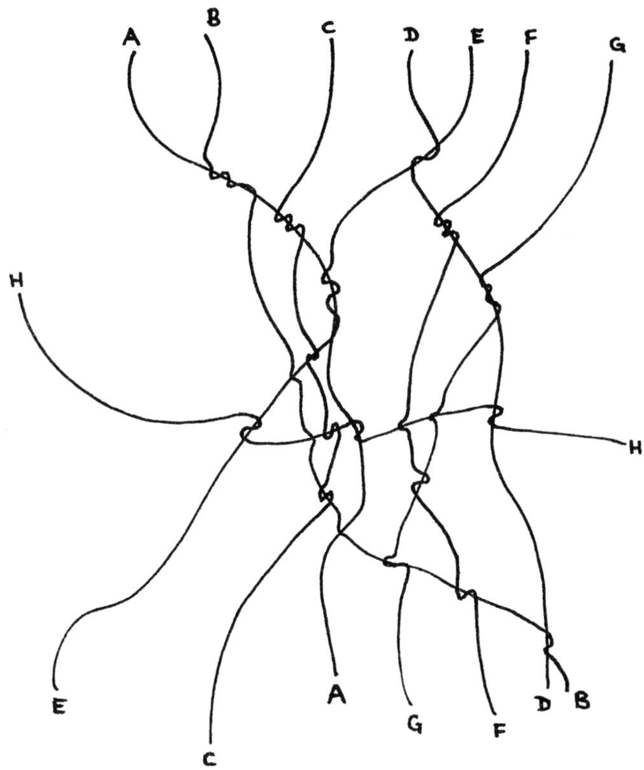

Figure 9.2. Meshwork, from *Lines* (Ingold 2007a: 82); by permission of the author.

think through and engage with biosocial life without the need to ascribe movements to some external force such as agency. Movement is always endogenous to the meshwork. Anything that might at first seem external to or separate from an organism, will, on a closer look, emerge as immanently related to the organism by some wayfaring line that meshes the apparently separate entities entwined through osmotic exchange and mutual sustenance. Humans, things and other forces in the meshwork do not interact as separate entities; they mesh, sustaining relations of contiguity and resemblance, and constantly overlapping with each other. With these considerations in mind, I now turn to life in the Enkopolwani sawmill.

AFTERNOON IN A SWAZI SAWMILL[54]

The Enkopolwani sawmill employs around one hundred workers, and runs two 8-hour work shifts.[55] It is located within an elliptical depression in the midst of a eucalyptus and pine tree forest (Figure 9.3). The first impression of the place as I approach along a muddy and extremely mushy driveway – the rainy season is no good for dirt roads – is of a noisy, dark space with smoke rising from a corner of the depression. A continuous, screeching metallic sound, the loud whir and chug of trucks, and whistling sounds as the trucks move back and forth, drown all other noises. I see piles upon piles of wood, and cannot decipher any particular order. There is a space marked by a tree, it marks out the car park, more by habit and practicality than by design. Nearer the entrance to this scoop in the ground are high piles of wooden planks, bundled and tied, looking ready for despatch to some timber market. They are sold in South Africa as an intermediate product: they are used to make cheap beds and other furniture; a famous soft drinks company uses them to build scaffolds for transporting cranes. The planks look wet and mouldy and I am surprised that they are not covered by polythene sheets. I wonder who would want to buy damp wood. Apart from the piles ready for shipping, the sawmill grounds are covered with finished wooden planks of various sizes scattered in unfinished piles of different heights, as well as wood off-cuts, bits of bark, broken planks, pine and eucalyptus logs, and small pieces of wood to be processed into mining poles. One cannot walk around the sawmill without stepping on wood. No clear pathways seem to have been created to prevent workers stationed by the machines from tripping over pieces of wood or other materials. Wet sawdust mixed with earth forms a squishy carpet of dark compost that fills the gaps still free of wood. Most of the machines are under a long shed of about 40 by 18 metres propped up by steel poles and covered with iron roof sheets. Walking under the shed is even more difficult (Figure 9.4); old mechanical saws and rusty conveyor belts are cluttered next to each other leaving hardly enough

[54] As a general rule, I will use the present tense in vivid descriptions aimed at rendering the fluidity and movement of real life as experienced in moments of fieldwork. This is not a timeless 'ethnographic present' of the kind rightly criticized by Fabian (1983), but an evocation of life in the making.

[55] About a quarter of them are women. Workers' numbers changed slightly during my fieldwork period due to high labour turnover. Up to April 2008, the workers were roughly divided in equal numbers between the two shifts. From April, the morning shift had about seventy people in it, and the afternoon shift thirty.

Figure 9.3. Panoramic view of the sawmill (courtesy of Bob Forrester).

Figure 9.4. Under the shed (courtesy of Bob Forrester).

space for workers, and far less space for wayfarers like managers, supervisors and anthropologists. There is no comfortable point of observation within this complex maze of materials. Running along the middle of the shed is a conveyor belt that moves planks processed by various machines to two cross-cutting tables where they are cut into specified lengths for the market. On the far end of the shed are several ramps where logs transported in the sawmill by tractors are piled high. At the end of two of these ramps are two Wood-Mizer machines (Figure 9.5).[56]

Wood-Mizers cut large logs longitudinally into two big slabs still bearing bark along the sides of their lengths. The off-cuts of this process are left in a messy pile at the side of each of these ramps. The slabs processed by the Wood-Mizer move along the conveyor to the multi-rip saw under the shed where four blades roll across the slab longitudinally, peeling away the left-over bark from both sides and cutting the central part of the slab into two big planks. These planks then move off the conveyor and are flung into a pile around a cross-cutting table. At the first table a worker is busy picking one plank at a time from the pile to cut to a specified size before flinging the cut piece to another pile a

[56] Wood-Mizer is a brand name. Some of the machines were named after their brand, especially when this was in evidence on the machine. Other machines were named according to the typology commonly used in the industry (such as the multi-rip saw or the double-edger).

Figure 9.5. Wood-Mizer and ramp (courtesy of Bob Forrester).

Figure 9.6. The Saw Kings (courtesy of Bob Forrester).

bit further ahead. Here a female packer moves, sorting out the planks into cubic stacks that will later be transported to the truck-loading area. Lengths of wood that are not thrown into the first cross-cutting table pile move further along the conveyor to two Saw Kings (also a brand name), where the planks are cut into even smaller planks (Figure 9.6).

Figure 9.7. The double-edger (courtesy of Bob Forrester).

But the Wood-Mizers are not the only apparent starting point for this rather messy assembly line. Under the shed, on the other side of the multi-rip saw, facing towards the big mass of wood off-cuts and sawdust that fuel an ever-burning fire of massive proportions, the double-edger machine performs the same function as the Wood-Mizers, but is used for smaller logs and requires a smaller number of movements by the machine operator, as it literally eats the wood up (Figure 9.7). Here too, bark off-cuts are thrown into a pile by the side of the machine. Sooner or later, each of these off-cut piles will come to life again, to be either burnt in the fire or reprocessed by the sawing machines. Processed planks from the double-edger move along a side conveyor to meet the multi-rip saw machine, where they are inserted into the same process as are those coming from the Wood-Mizers. Both types of planks – or to be precise, some of their parts – will at some point reach the Saw Kings. After being eaten up by the Saw Kings, the smaller timber planks are flung into a pile adjacent to the second cross-cutting table, where they are cut into lengths by production specification. These are then sorted by packers, most of them women.

Although some women work with the machines, most are in the packing section (Figure 9.8). They often stop and cross each other's paths to talk and exchange jokes, gossip and chat, so that their bodies can slow down and regain some energy. Wood planks are sorted by size, that is, length, width and thickness, and by quality – specifically, in two grades, 'grade A' and 'grade B'. The sizes of the planks are more

Figure 9.8. The packing area (courtesy of Bob Forrester).

obvious and easier to determine, even to the untutored eye, than the quality of the wood. Five packers, all women, are busy sorting and packing the wood around the second cross-cutting table, outside the reach of the shed which stops where the two Saw-Kings are. It seems that the women are moving in random elliptical lines among different scattered points, marked by the presence of unfinished and unlabelled rectangular piles of planks. All the same, they know where to put the right planks.

As I move away from the packing area, I am curious to discover where the holes in the ground of the covered area lead. I see workers at different work stations throwing wood off-cuts there, as there are holes near most of the machines (Figure 9.9). I discover that the waste goes down along chutes that lead to conveyors, manned by two workers whose job is to make sure that the mixture of sawdust and wood pieces does not stop in its movement towards the outside (Figure 9.10). The workers are covered with the same materials that move along the conveyors. Caught by my Marxist prejudice, I see this as a metaphorical transposition of hell, with people underground somehow 'punished' for some sins they must have committed. Ironically enough, the waste that travels through this section will end up in the big fire about 20 metres away in the open.

Like every other time I visit the sawmill, after a few minutes of aimless wandering around and trying to make sense of what is going on, I resignedly head towards the edge of the packing area near the big tree where I parked my car. I sit on one of the unfinished piles,

Figure 9.9. One of the holes leading to the waste area underneath (courtesy of Bob Forrester).

Figure 9.10. Working under the sawmill platform (courtesy of Bob Forrester).

momentarily sheltered from the helter-skelter of the sawmill. The screeching noise of the Saw Kings is the background against which my attempts at capturing the relentless outpouring of life end up in a clutter of half-finished thoughts. I open the hard A5 notebook which I firmly held in my right hand as I walked through the sawmill; I look at my notes. As I sit on the relatively comfortable improvised bench, sustained by a vertical accumulation of planks, I pause once again to look at the muddy ground: an irregular overlap of footsteps, small and big wheeled tracks left by cars and trucks. It is a damp, grey summer afternoon with occasional drizzle. The rainy season is well on its way and it rained heavily yesterday night. The ground nurtures other forms of life too. I can already feel the fleas tickling my ankles.

LOOKING FOR THE ASSEMBLY LINE: THE ILLUSION OF ORDER

Among all the fieldwork activities I carried out in Enkopolwani, my regular visits to the sawmill were by far the most challenging. The sawmill was the one place where any hint of certainty and academic 'control' would quickly disappear. Trying to emulate many social scientists before me (Beynon 1973, King 1978, Linhart 1981, Thompson 1983), day after day I kept looking for the semblance of an assembly line. What were the stages of production? How were timber logs transformed into planks? How did workers and managers struggle along the production line? These were some of the questions that I was grappling with at the time.

With the help of my wife, Patience, we came up with a two-dimensional floor plan of the sawmill (Figure 9.11). Patience found it easy to draw abstract orthogonal lines to mark the different workstations and the specific locations of the machines along my yearned-for, imaginary assembly line. The problems started when we tried to insert people into the picture. During the course of a whole eight-hour shift it was sometimes the case that every machine might have been used at some point or another, but I never once witnessed the sawmill working at its full capacity. Except for short periods of intense activity, for the most part production had a stop-go rhythm with materials quickly piling up at one workstation while running short at others. Some workers would move from one workstation to another, while others were idle due to blockages, breakages, temporary slow-downs, or lack of input materials to work with. Still obsessed with ambitions of completeness, we attempted to juxtapose on the sawmill floor-plans

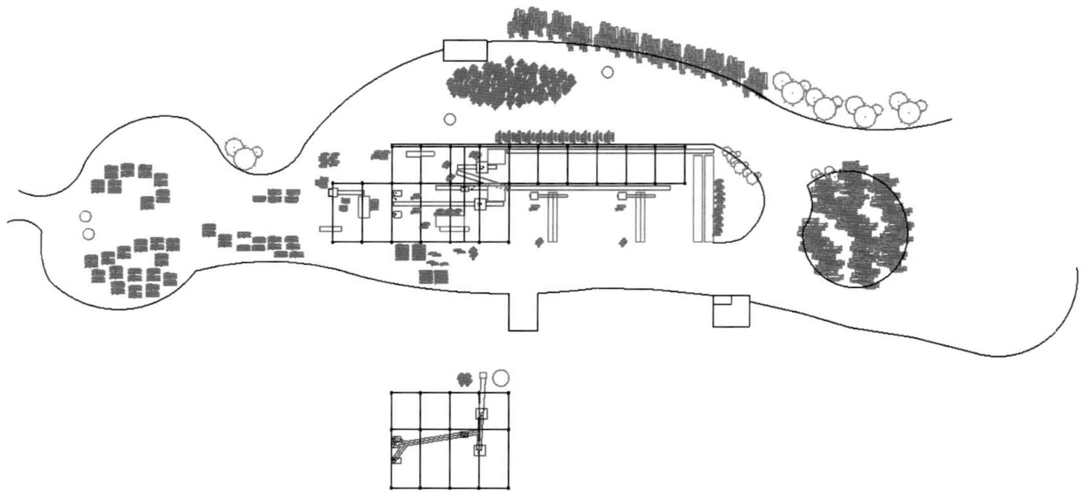

Figure 9.11. Overview of AutoCAD sawmill floor plan.

workers working at different times during one shift or even in different shifts. In Figure 9.12 you can see a blown-up section of what we came up with, showing the interactions between workers, machines and wood at the Wood-Mizer workstations and the linkages between the Wood-Mizers and the multi-rip saw. We ended up, once again, with points and straight lines. No matter how skilled or ingenious the charts might look, no life was flowing through them. The exercise was highly instructive, as it clearly highlighted the limitations of the network logic.

WAYFARING AND THE AFFORDANCES OF THE ENVIRONMENT

Despite my aimless and irregular wanderings, people at the sawmill soon got used to me, as a familiar addition to the meshy landscape of half-finished planks, off-cuts, workers in blue uniforms, mud, sawdust and a variety of other materials. The activity of walking around without a pre-established route and with no limitation to any specific area made my wayfaring similar to that of Mduduzi, the sawmill line manager, and the shift supervisors. Just as wayfaring lines do not have a beginning or an end, but trace movements along, so wayfarers do not have a specified goal that rigorously dictates and constrains their actions. The work of Mduduzi and the shift supervisors consisted mainly in finding their way through the chaotic meshing flows of materials to get a hunch of where production could be increased and blockages undone. The aim, quite predictably, was to increase the number of finished planks that ended up in the relatively neatly packed piles adjacent to the truck-loading area. Yet they did not work with mental maps and a standardized set of instructions. They had to feel their way through. They spotted the areas where too many workers were concentrated without any visible positive effect on the speed of the tasks carried out, or where too few workers were left on their own while others stood idly in the irregular and poorly demarcated interstices among different workstations.

What influenced the movements of humans, things and materials were not so much a priori notions of class struggle and worker-management conflict, but rather what Gibson (1979: 43) calls 'affordances'. Affordances are properties of the environment that constrain certain flows and movements and facilitate others. In this case, what is commonly referred to in the literature as 'shirking' (Parry 1999: 126), can also be conceptualized as a human adaptation to the constraints

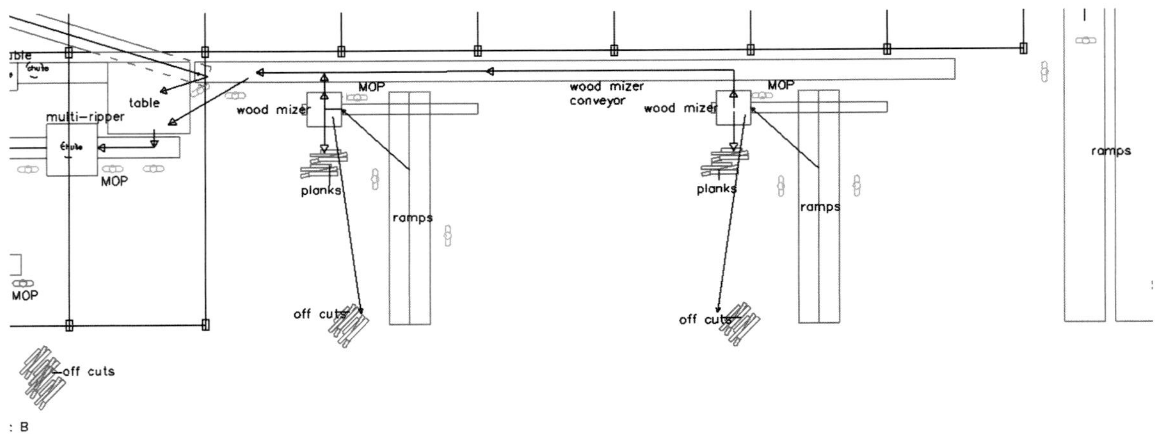

Figure 9.12. Enlarged section of sawmill floor plans with human–machine interactions; the circles represent workers.

and potentialities offered by the environment. The lack of clear demarcations between one workstation and another, and the resulting overcrowding of machines, off-cuts and mud, afforded very little mobility to the workers in their attempts to harness the flows of materials coming their way. This could even prove near-lethal, as large pieces of wood occasionally bounced off the machines due to bad maintenance of the blades or other blockages. At the same time, it made supervision and managerial wayfaring extremely difficult, whilst easing workers' attempts to get some rest or to reduce their physical contribution to the work tasks. There was no vantage point around the sawmill where one could gain a bird's eye view of the production process. Workers and supervisors moved along wayfaring lines, which met and crossed each other and formed irregular patterns that could not be captured by rectilinear representations.

FROM INTERACTIONS BETWEEN POINTS TO TENSIONS IN THE KNOT

As my wayfaring mind wanders around the sawmill, I see materials continuously transformed into other materials; surfaces rapidly produced and dissolved through manifold ways of cutting; pieces of wood giving way to other, differently shaped, pieces of wood; human hands inserting themselves sometimes on the rough surface of wood, sometimes on the smoother metal surfaces of old sawing machinery. Nor do these processes follow any prescribed or repeatable order. As I track a piece of wood while it moves through the different work stations, it is not the same thing ever again, at any point in time. It is continually being transformed into something else, sometimes even separated into different things. It can be a log that is transformed into two major planks; the two planks are now separated and will be transformed into even more planks of different sizes and lengths. The bark of the log, once attached to the same body of wood, is now removed from it, adjacent to other barks previously detached from other bodies of wood I arbitrarily call 'logs'.

At the interface of log, Wood-Mizer machine and machine operator, there is no real separation between subject and object, or between object and object (Figure 9.13). Is the machine cutting the log into pieces, or is the log going through the machine, transformed by the complex succession of single movements, of which the machine operator is also an inseparable part? Is the machine operator thinking of himself as a separate agent controlling wood matter and mastering the

Figure 9.13. Wood-Mizer in operation (courtesy of Bob Forrester).

might of the machine? Or, more realistically, is he busy trying not to get injured, as he knows that the elliptical steel blade can snap at any time, quickly enough to decapitate him? How can one assert the absolute independence and clear-cut separation of machine, log and human if not by a leap of the imagination? Surely, an awareness of the log is not just a figment of the imagination, but the result of years of training through which one learns to perceive, separate and categorize. But in practice, there is no thinking separate from the action and the materials involved. As materials permeate from thing to thing, from bundle of wayfaring lines to bundle of wayfaring lines, movement and flux is all that can be seen, perceived and imagined. When one focuses on movement, the clearly demarcated points and straight lines constituting machines and workstations in Figure 9.11 dissolve into knot-like bundles whose constituent threads go their separate ways beyond the knots themselves. The abstract rectilinear forces of production and their internal contradictions, called upon by the Marxist factory ethnographer to explain the causality and intentionality of life on the shopfloor, are here replaced by the tensions produced by the knotting of life, devoid of starting and end points (Ingold 2011: 218).

The workstation can only be seen as a thing in so far as flows of materials move in and out through its porous boundaries. The boundaries are nothing but the illusion of approximation to a clearly demarcated object afforded by the intensified meshing of lines in certain

Figure 9.14. Flinging plank near the cross-cutting tables (courtesy of Bob Forrester).

areas of the environment. Look, for instance, at Figure 9.14. The two straight lines of the cross-cutting tables intersecting each other provide a short-hand for mapping. Yet, there is much more going on here. Planks are flowing in from the Saw-Kings, the metal blue sawing machines just opposite the tables, near the covered area. One or two move directly to the cross-cutting tables, but the rest are piled up around it. The sawdust expelled from the Saw-Kings also accumulates in irregular heaps, which slowly expand towards the metal legs of the cross-cutting tables. As the operator by the table cuts one plank, an off-cut on the right side of the table is already about to detach itself (Figure 9.15). At the same time, a plank is flying in (or flung into) the air, about to land between one of the five poles stuck in the ground, where other planks are waiting to be picked and transported to the packing area (Figure 9.14). What makes the cross-cutting area into a 'thing' is the confluence of multiple flows of materials into a bundle of entangled lines. The pictures also clearly show how the so-called 'product' is another fiction, a Platonic ideal form that materializes into dozens of different pieces of wood, only ever approximating, if at all, to the abstract model. And even as they are named 'pieces of wood', they are not separated from the flows of materials that constitute them. The flying or flung plank is very much a plank-in-the-air, the off-cuts near the Saw-Kings are hard-wood-in-the-sawdust, and the planks grouped in between the wooden poles are more correctly planks-on-each-other. This 'thinginess' of things (Ingold 2010b: 96) is in itself a

Figure 9.15. Cross-cutting table (courtesy of Bob Forrester).

property of materials meshing with other materials, coming to life with their own properties, together shaping certain paths of becoming while occluding or inhibiting others.

THE SOCIAL IN THE BIOSOCIAL: IRONY AND DANGER

So far the discussion has focused primarily on what are conventionally referred to as 'physical' or 'technical' properties of the environment. If the physical is indeed social, or better biosocial, can the social really be biosocial? Are there no a priori distinctions between matter and mind, the social and the material? To provide some partial answers, I will now turn to the enmeshing of irony and danger.

Humour and laughter, in varying degrees of subtlety and bluntness, were a constant of life in the Enkopolwani sawmill. Workers often told each other jokes and funny stories, both while working and during lunch breaks. More often than not, the irony meshed with other wayfaring lines of wood production. These were not just jokes 'about' something; they were narrated as workers went about their physical tasks. Irony became a widespread property of the environment, no less important than other 'technical' activities like, for instance, cutting through wood.

On one occasion, Sicelo, a young male assistant operator, is busy rolling a log from the ramp to the Wood-Mizer machine. This task is particularly hard and a worker risks serious injuries to the hand, fingers and wrist. It takes a particular mix of strength and dexterity to perform this job safely. Muscular power needs to be applied in conjunction with a specific movement of the hand that releases the log from the pile.

As he works, Sicelo sings a tune by Malaika, a popular South African band. The song in question is the disco version of 'Never change my mind', often heard in bars and clubs. The song lyrics are in Zulu, except for one line in English referring to the title. Zulu is close enough to SiSwati for SiSwati speakers to understand it easily. Sicelo is singing the first verse of the song:

Noma ngingafa	Even if I should die
ngifela nin' abantwa bami	I would die for you my children
Ngeke mina ngihlukane nani	Never will I part from you
Ningabantwana bakaBaba	you are the children of the (Heavenly) Father

As he sings, a log rolls down the ramp faster than anticipated. Fortunately no harm is done and the log lands safely by the Wood-Mizer. It is just another of the manifold situations in the everyday routine of the sawmill where a work accident could have easily taken place. After the danger is over, Sicelo speaks again:

> *Noma letigodvo tingangigcoba ematsambo, ngentela bantfwa bami.* – Even if the logs break my bones, I'm doing this for my children.

As he speaks, he poses towards his audience – the other assistant operator, the machine operator, me and my research collaborator – in mockery. By reference to the song, the sentences above are metaphorical substitutes for the first two lines of the song verse:

> *Noma ngingafa, ngifela nin'abantwa bami.* – Even if I should die, I would die for you my children.[57]

A number of wayfaring lines of imaginary materials are entangled here with the more 'physical' lines of wooden and metal materials and of Sicelo's bodily movements. The metaphor brings together in the same space the Christian discourse of Jesus dying on the cross to rescue the believers (his 'children') from sin, Sicelo's struggle to get by and feed

[57] Note that this is Zulu, while Sicelo's ironic words are in SiSwati. Hence *abantwa* is the Zulu orthography for the SiSwati *bantfwa*.

his children on a meagre salary, and the constant dangers posed by his daily involvement in the production process.

Within the field of linguistic anthropology, metaphor is conventionally defined as a logical operator that links together two or more previously separate domains (Fernandez 1974, Basso 1976, Sapir 1977). In these terms, metaphor is still embedded in a rectilinear framework of point-to-point connectors. What is happening here, however, is not the interconnection of supposedly distinct domains of religion, family and the workplace. Rather, metaphor makes it possible to trace movements along the several wayfaring lines issuing from the particular knot of the Wood-Mizer ramp and extending all the way to other bundles, making up the movements and tensions of family life and supernatural forces. As different knots of life are tied together through wayfaring lines, new knots are created and juxtaposed with the old ones. Sicelo's lines of family life, his struggle to feed his children and fulfil his aspirations as an adult man and father, mesh with the other materials of the sawmill. The lines reaching out and beyond the mill are also always within it. Because there are no a priori categorical spatial or temporal distinctions among the different kinds of materials meshing together, the environment is posited here as an unbounded epistemic space. It cannot be mapped out in its limitless totality, but only explored by moving along the lines that constitute it. In this sense, the meshwork is not an abstract representation of a space; it is a perceptual tool to aid in wayfaring through the environment.

In the same way, movement cannot be represented by cause-and-effect linkages and law-like regularities; it is the result of the continuous repetition-cum-variation produced by the tensing and meshing of flows of materials. This movement is an improvisation. The term 'improvisation' effectively dissolves yet another dualism, that between repetition and variation. Nothing is ever repeated, yet things are produced from the momentary meshing of other pre-existing materials. To move along the flows of materials, Sicelo needs to improvise. He improvises to feel the right point at which the timber log will move with the least amount of effort, without rolling down the ramp at a dangerous speed. He improvises to deal with the blockages which he regularly encounters as part of the often hostile flows of life in the sawmill. In Ingold's words, 'to improvise is to follow the ways of the world, as they open up, rather than to recover a chain of connections, from an end-point to a starting-point, on a route already travelled' (2010b: 97).

To make a joke is another way to improvise, to discover and create new wayfaring lines to ease the tensions produced by other, less favourable, ones. The joke is not an abstract commentary over a detached object of conversation. Sicelo sings as he works, the two activities are concurrent. He sings and hears himself singing, as his hands struggle to pluck a log out of the pile, feeling the touch of the rough wood through the mediation of his protective gloves. As the log rolls out of control, there is silence. The joke comes after a few seconds, yet it is a continuation of the same long moment. The bodily experiences of danger, fear and alertness provide the materials for the metaphor.

The social commentary is embodied in the materials of the moment. The expression 'even if the logs break my bones' has a very literal meaning and refers to the physical strain caused by the tasks Sicelo performs every day; but it also indicates the imagined and perceived danger that one day, in the flash of a second, the everyday physical struggle might turn into outright tragedy. Comic relief and metaphorical play help Sicelo and his audience to counteract the tension of the danger just passed, and to reinsert themselves into the flows of life from which they have been momentarily displaced. On a practical level, the joke buys Sicelo time to regain composure and get ready for his next log, with the overt aim to minimize damage to himself and his co-workers. It also helps him to make sense of a potentially harmful event and to get on with the task ahead. Sense-making, here, is not a form of disembodied reflection about the external world. It is an improvisatory practice that occurs from within the environment, a perceptual practice of being-in-the-world.

STRAIGHT LINES OF CAPITAL?

At this point, our imaginary Marxist factory ethnographer might still object that there is a bottom line that no wayfaring lines, flows or other epistemological wizardry can transcend: capitalist production only exists because surplus value is extracted by exploiting workers' labour. He would then go on to say that this holds true in the Enkopolwani sawmill as in anywhere else. How can a meshwork perspective accommodate this 'fact of life'?

That the sawmill employed cheap labour in very poor and dangerous working conditions – even by local market standards – is indeed a fact.[58] That it produced profit was less straightforward than might at first appear. The intention of EMS managers was certainly for the

[58] The vast majority of workers in the sawmill and other Enkopolwani economic activities earned between 35 and 40 emalangeni per day. In 2008, 35 emalangeni

sawmill to run at a profit so that in the long term this would contribute to make the town and its orphan social services self-sufficient, that is, not reliant on external donor funding as they were at the time. However, there was no clear budgeting process in place to establish if the sawmill was running at a loss or not, nor did the managers seem too bothered about the mill's profitability. A closer look at the 'woodiness' of Enkopolwani reveals a story that sits uneasily with the rectilinear assumption that the extraction of surplus value from workers is the main driver of production.

WOOD EVERYWHERE

I have already shown how the sawmill is filled by all kinds of wood materials. Well beyond the confines of the sawmill, most of the Enkopolwani private property is covered by pine and eucalyptus trees. Off-cuts from the sawmill are transported to the workers' residences, where they are lit in old wood-stoves, used to cook and to warm up the houses during the cold winters. People also fetch wood from the forests on their own, despite EMS management warning that such activities would be considered 'theft'. Workers employed by timber contractors often cut logs for a small fee, this too against the best advice of management. Tractors with trolleys full of timber logs are seen every day moving up and down Enkopolwani's badly maintained main tar road that runs between the town buildings. Branches and pieces of bark, as well as small and bigger logs, are often visible on the tar, as one drives or walks through it. The polished wood furniture of the lodge is also made from planks cut at the sawmill, and then further transformed by the workers in the carpentry shop. The forests were planted around the time of independence from Britain in 1968. Back then, a small sawmill produced timber poles, and wood for the workers' domestic needs. One way or another, in Enkopolwani 'woodiness' is an inescapable fact of life.

Nor did it escape Stephen, a private businessman who bought Enkopolwani in 2004, three years after the liquidator had put the property on the market. The price of sale was a real bargain, not least because there were no other willing buyers. The liquidator had approached two

were roughly equivalent to 2.20 British pounds. Although this was considerably above the minimum wage set by government for the forestry industry (18 emalangeni per day, plus food rations or a cash equivalent), the salary was considerably lower than that paid by the two major multinational timber operators in Swaziland, where minimum wages in 2008 were set at or above 70 emalangeni per day. Both in Enkopolwani and in the workers' compounds of the multinational competitors, workers did not pay for accommodation.

timber multinationals with major operations in Swaziland. They were not interested because they could not run a viable operation. Their projected operating costs would have been too high due to higher salary packages and collective bargaining with recognized unions, high-tech state-of-the-art machinery and the requirement to adhere to health and safety regulations. The low-quality wood and the badly maintained forests would not have justified the investment. Stephen had recently left his reputable and profitable professional practice to pursue his dream to run a 'real business'. He did not quite know what he was getting into, but he did it. He bought Enkopolwani. He now owned a ghost town. One of his previous clients was Dan, an independent timber operator running several sawmills throughout Swaziland. Improvising along the flows of life, Stephen started the tourist lodge and brought in an excellent wood carver to produce high-quality arts and crafts. The bulky traces of the previous mining operations – prominent mine dumps, the cableway leading into South Africa and the adjacent workshop area – would have become a museum about the history of the Enkopolwani mine and about mining in Swaziland.

Dan teamed up with Mduduzi, a security guard who attracted the attention of Stephen. Stephen thought he would make a good manager, as he was well respected in the neighbouring customary land area, and showed promise in the complex art of managing local labour. With a dozen men and a few pieces of old sawing machinery, the shopfloor was set up on the concrete platform left behind after old power station machines and other scrap materials were sold during the liquidation period. The sawmill grew out of an organic assemblage of materials adapting to the affordances of the left-overs of the power station. In a couple of years it looked pretty much as it did in 2008. The sawmill brought in cash, as Dan integrated its business operations into the marketing and sales function of his other sawmills. The other projects did not take off, though. Stephen needed more money to finance his emerging vision of a major business development centred around cultural tourism and arts and crafts. A multimillion dollar mine rehabilitation grant from a high-profile international monetary institution never materialized after complex negotiations with different echelons of government failed, leaving Stephen strapped for cash.

SOME WOOD, SOMEWHERE

It is now early 2006. Dan owns half the shares. The sawmill keeps on growing. Stephen is looking for a way out. Dan finds some enthusiastic buyers. Adrian, a Canadian businessman and a fervent Pentecostal Christian, is eager to lend his business skills to the service of the poor. After exploring some charity projects in Germany and Australia, he finds his way to Swaziland. This is the place God has chosen for him. Dan's long-time family friend Marc is also 'born again', and eager to embrace a new life outside commercial business. Dan and Marc grew up in the south of Swaziland, in the tight-knit circle of Afrikaner farmers whose families had been there for generations. It turns out that Dan, too, is a born-again Christian. Their spiritual leader is Pastor Alan, the son of an Irish man who came to Swaziland to find his fortune in the late colonial period; Alan is a charismatic Pentecostal pastor who runs a successful retreat for street kids near the capital.

Dan, Marc, Alan and Adrian meet. The Enkopolwani vision comes together. EMS buys out Stephen. Dan donates his shares. The sawmill is still running. The forests are still there. Donor funding starts to flow in to finance orphan care and other social services. Running expenses are high. The sawmill remains integrated in Dan's private businesses and keeps on bringing in much-needed cash. Dan is one of the founders of EMS. What started as a pure profit-making activity now gives him prestige and status in the wider white community. He is helping out EMS by generously providing all the services of his business infrastructure for free. The planks are loaded onto his trucks directed to a South African depot, just as before. Plans change, old business partners go and new donors come, missionaries and volunteers grow in number, workers come and go, sometimes staying for a few months, some even a few years. And yet every day, as the sun rises over the hills, the only certainty is that some piece of wood somewhere in Enkopolwani will come to life, at least until a major fire like the ones that burnt down most of the neighbouring forests in July 2007 leaks through. And even then, it would take some time to clear the grounds.[59]

Concluding remarks (and other beginnings)

I have shown how the meshwork can be employed as a productive tool to navigate through the complex realities of a Swazi sawmill, and to

[59] In July 2007, major fires burnt vast areas of forests, affecting many timber plantations in the western part of Swaziland. While the neighbouring forests were heavily damaged, residents managed, with a sustained collective effort, to keep the fire out of the Enkopolwani perimeter.

move away from the rectilinear logics and assumptions of the network, here sketched out in reference to Marxist factory ethnographies. The meshwork does not dragoon the reality under study into neatly ordered analytical statements. Instead, it sensitizes the anthropologist's – and the reader's – perception to the flux and movement of real-life becomings. The ethnographic journey presented here started in the middle of an elaborate and often impenetrable mesh of different organisms and materials. The purpose was to highlight patterns of relations, like wayfaring through the irregular paths of the sawmill grounds, rather than an abstract map of relata of the kind presented in Figures 9.11 and 9.12. As the narration developed into further detail, I challenged the dualisms of mind and body, social and biological, human and machine, through a detailed description of the dynamics at work in the enmeshing of human, machine and wood materials in the sawmill workstations. Just as it is not productive to consider the physical in isolation from its biosocial entanglements, it is also important to avoid the a priori division of the social from the environment in which it is enmeshed. This is what I argued with Sicelo's story, where the mesh of irony and danger in the Enkopolwani environment shows once again that becomings are always biosocial. In the final part, I pushed further the scaleable and permeable boundaries of the sawmill and its wider environment – impermanently located both 'within' and 'beyond'. A consideration of the intricate micro-histories behind the 'woodiness' of Enkopolwani helped deconstruct another universalist assumption of Marxist factory ethnography, that of the managerial extraction of surplus from workers in capitalist production. While the stated goal of EMS managers was to make timber production profitable, this logic did not explain why the sawmill started in the first place or how wood production and its overlapping realities came to play such a central role in the everyday life of Enkopolwani.

Is the meshwork, then, just another theoretical closure, perhaps excessively influenced by exogenous knowledge that is superimposed upon the local realities under study? Is it merely a better substitute to remedy some of the explanatory shortcomings of the specific Marxist framework presented here? Rather than a set of abstract statements, the meshwork acts as a sort of vaguely visible plot that binds together the threads along which the everyday epic of life unfolds with its often eccentric protagonists – human and non-human organisms, things and forces involved in a continuum of overlapping and differentiating movements in one and the same unbounded totality. To use a

Bakhtinian metaphor, the language of the meshwork is characterized by heteroglossia, that is, a multiplicity of overlapping and often contradictory registers and speeches unfolding as a stratification internal to language (Bakhtin 1981: 271–272), rather than an interaction of separate and distinct voices and languages. In this sense, the meshwork is endogenous and complementary to the life processes evoked. Nor should this heteroglossia be read as a synonym for a naïve and unproblematic togetherness. As Sicelo's open-ended parody of the precariousness of life poignantly indicates, contradictions, obscurities, and unresolved tensions cannot be disentangled from the constant meshing of biosocial materials. In the interstices of the knots resides yet more life to be explored, some other time, by other wayfarers. The adventurous traveller is warned though: expectations of 'true sight' and mastery of the unknown are likely to be disappointed. All one might be left with is a burdensome awareness of the inextricable entanglement of the visible and the invisible (Merleau-Ponty 1968: 130–131).

ACKNOWLEDGEMENTS

This research would not have been possible without generous support from the Economic and Social Research Council, award no. PTA-031-2005-00040. Mlungisi Dlamini provided invaluable editorial and linguistic assistance; he edited an earlier version of this chapter, provided a translation from Zulu of the verses from the song 'Never Change My Mind' by Malaika, and checked all SiSwati and Zulu spelling and translations in the text. Mabandla 'Wiseman' Dlamini, Mandla Shabangu, Nathi Motsa and Mncedisi 'Spice' Zwane proved indefatigable field collaborators, and helped with observations, cultural analysis and linguistic translations. Wiseman's fieldwork notes were key to some of the ethnographic descriptions presented here. We are also grateful to Harri Englund, Gisli Palsson, Tim Ingold, Christopher House, Andrea Grant, Mantas Kvedaravicius, Henrietta Moore, Koen Stroeken and Wim Van Daele, for reading and commenting upon earlier drafts of this chapter. Informed verbal consent was secured from all photographed subjects appearing here.

ISTVAN PRAET

10

Humanity and life as the perpetual maintenance of specific efforts: a reappraisal of animism

THE ODDITY OF MODERN CONCEPTIONS OF HUMANITY AND LIFE

One way of describing contemporary western cosmology would be to say that it is characterized by notions of humanity and life that are curiously unconditional. Western societies are often supposed to be exceedingly bureaucratic, yet they display an astonishing *lack* of bureaucracy in this particular respect. The requirements that have to be fulfilled to be counted as 'human' or 'alive' are absolutely minimal. No forms have to be filled out, there is no waiting list or exam to be passed, no proof of competence is needed, no trial period has to be completed. Provided you are born of a human mother, no specific effort has to be made whatsoever: the moment you come into the world, you are human (as well as alive) by default – no further questions asked. As long as you keep breathing, nobody will challenge your status as a 'human being'. In this context, humanity is entirely maintenance-free. The same applies, more generally, to life. As long as you are animate, your status as a 'living being' is assured. Within modern, western cosmology both humanity and life are given, in the sense that they are not something you have actively to work towards and deliberately need to uphold at all times. In such a framework, to ask whether something is human or alive is a matter of either/or rather than degree (even though this axiom is beginning to be questioned in such fields as cell biology; Landecker 2003). In this chapter I demonstrate that this widespread attitude of unrecognized complacency is

Biosocial Becomings: Integrating Social and Biological Anthropology, eds T. Ingold and G. Palsson. Published by Cambridge University Press.

actually quite strange. This becomes apparent if one considers notions of humanity and life more broadly, especially from the perspective of so-called animistic people. In what follows I illustrate through ethnography that if there is one thing that unifies the most diverse forms of animism across the world, it is a consistent emphasis on achievement rather than birthright. My suggestion is that animistic notions of humanity and life share a highly original yet poorly understood feature: they depend on continual, never-ending efforts. Such efforts are neither rigidly constricted nor entirely random; even though they are to a certain degree negotiable and may evolve over time, they are fairly narrowly circumscribed and relatively stable. In such a framework, I aim to show, humanity and life are always conditional. Within animism, both humanity and life entail specific efforts and require perpetual maintenance.

Another way of describing western cosmology would be to say that it is characterized by notions of humanity and life that are peculiarly inclusive. Consider the remarkable ease with which practitioners of modern science attribute the status of 'human being' as well as that of 'living being' to this, that and the other. In contemporary anthropology even the strangest outlanders are unhesitatingly considered human. Suppose you are a British anthropologist studying the French. No British person will contest that the French are blatantly exotic. But despite this commonly acknowledged exoticism, no anthropologist would ever doubt the existence of a common denominator between the British and the French, namely humanity. There may be a significant difference, but that difference is never conceived of as total. And what goes for the anthropological notion of humanity also goes for the biological notion of life. In current biology even the most improbable, outlandish entities are attributed the status of 'living being' without a second thought. The way in which marine biologists envisage the unfamiliar universe of deep-sea microbes is a case in point. Microbial extremophiles thrive in extreme surroundings such as high-pressure, high-temperature volcanic vents on the seafloor (Helmreich 2009). Given that no biologist contests the spectacularly alien nature of such microbes it is curious that one obvious question is consistently avoided: nobody wonders whether they should be considered alive to begin with. The microbes' inclusion within the realm of life is deemed obvious. The possibility that we are dealing with a novel domain of non-life is inconceivable. In a similar vein, the French are designated as foreigners or 'other humans' – the idea that we are dealing with non-humans remains unthinkable. In the biological worldview life acts as a powerful magnet that draws all animate beings into its jurisdiction; in the anthropological worldview humanity acts as an equally powerful magnet that attracts all specimens of what scientists refer to as *Homo sapiens*. In this chapter I aim to make clear that these inclusive notions of humanity and life are actually far from self-evident by focusing, again, on conceptions of humanity and life among so-called animistic people. My central argument is that such 'animistic' conceptions are – or at least used to be – impressively uniform all over the world in so far that they are always conditional and (hence) restricted. Whether we are talking about humanity or life, nobody is ever included unconditionally. To be human and to be alive always requires the maintenance of a specific effort.

THE SURPRISING CONSISTENCY OF ANIMISTIC NOTIONS OF HUMANITY

My project is comparative in nature, and is based on ethnographic material not only from lowland South America (my area of specialization), but also from North America, Asia and Africa. However, in this chapter I will focus almost exclusively on one particular indigenous group, the Chachi. The reason why these Amerindian inhabitants of the Pacific coast of Ecuador are accorded pride of place is that I know them first-hand. Between 2004 and the present I have conducted 18 months of fieldwork in the coastal lowlands of Ecuador. And I would like to begin with something that happened in the early stages of my fieldwork. One evening, the Chachi school-inspector Santiago Añapa Pianchiche looked up from his newspaper and told me:

> Can you imagine, according to this report there are now six billion Chachi in the world!

We were sitting in his house in *el tercer piso*, one of the poorer neighbourhoods of Esmeraldas, a small town on the Pacific coast of Ecuador. What struck me in this phrase was not so much the estimation of some demographer cited in the newspaper, as Santiago's curious use of the term 'Chachi'. Until then, I had only heard people using it with regard to the specific Amerindian group of approximately nine thousand people among whom I was doing ethnographic research. I had always believed that 'Chachi' was simply the name of a particular kind of 'indigenous people'. Noticing my confusion, Santiago explained that it can be used in a wider sense as well. All the people in the world, he insisted, could be considered 'Chachi':

> Tsachila, Epera, Blacks, Quichua, Hispanics, Gringos ... all are Chachi. It is not just the Chachi in our own community that can be designated as such.

At first glance, Santiago's remarks confirm the universality of the concept of an incontrovertible, unified humanity. Indeed, what people such as the Chachi say seems to establish the 'obviousness' of the anthropological notion of a singular human species rather firmly. With regard to humanity, westerners and those traditionally portrayed as non-westerners – that is, anthropologists and so-called indigenous people respectively – appear to march in perfect unison. But can one assume that this has always been the case? Santiago for one was quick to add that this specific usage of 'Chachi' is fairly recent. In his view, it coincides with the emergence of the bilingual education system during the 1970s. In *Cha'palaa*, the vernacular language, there is no original term for 'the human species' as conceived of in modern anthropology. The notion only emerged when people began to learn Spanish. Santiago, himself a pioneering bilingual teacher, then explained how 'Chachi' was used in the past:

> Formerly, 'Chachi' had a very narrow meaning; it uniquely referred to those who shared our language and lived like us. Quichua Indians from the highlands, blacks from the coast and city-dwelling Hispanics were not considered Chachi. In fact, the latter were not considered proper humans. 'Chachi' literally means 'humans' [*gente* in local Spanish] or 'true humans' [*gente verdadero*]. Therefore, when we said that somebody was not Chachi, this amounted to saying that he or she was not really human. Moreover, you should not think that this specific usage is merely something of the past: along the River Cayapas and elsewhere in the countryside it is still quite widespread, even among those who regularly visit cities.

This particular conception of humanity, which Santiago described so succinctly, is by no means unique to the Chachi. South American Indians rather consistently refer to themselves with terms that are often translated as 'the people', 'the real people', 'true persons', or 'us, the human beings', implying that those unlike 'us' are not truly human. Lévi-Strauss described this phenomenon long ago; in *Race et histoire* (1975) he already suggested that the notion of a humanity encompassing all forms of the human species was entirely absent for millennia. Humanity, he pointed out, stops at the boundaries of the tribe, the linguistic group, sometimes even the village.[60] In the anthropological literature this is often referred to as the ethnocentric attitude of small-scale populations. What is remarkable, but for some reason hardly ever emphasized in contemporary studies, is its worldwide occurrence. Table 10.1 may only be a crude and far from exhaustive illustration, but at least it has the merit of drawing attention to how widespread this notion of a restricted humanity really is, or at least used to be:

Of course, drawing up an overview such as this one is somewhat hazardous. The way in which so-called indigenous people are designated is a notoriously thorny topic. In many cases, the etymology of the terms of self-reference is convoluted and the people concerned disagree on their exact meaning. In addition, such terms are by no means static – think of my initial example, Santiago's account of the history of the expression 'Chachi' (Erikson 2004). The fact that early travellers and ethnographers have often taken specific ethnonyms at face value has certainly added to the confusion. It is now well-known that many of them have designated the people they encountered with derogatory terms, because they simply adopted the names used by their neighbours or enemies without much verification. Even if one takes great care to select only the most reliable and up-to-date sources (as I have tried to do to the best of my ability), the matter remains tricky. One could ask, for example, whether some of the translations I have listed (such as 'real people', 'true people') are entirely adequate, given that many contemporary ethnographers question whether notions such as 'reality' or 'truth' are universal. One could also take issue with the rather haphazard geographical distribution of my examples, a consequence of the limited availability and the scattered nature of relevant ethnographic material. The fact that 'Europe' as an ethnographic region has been omitted from the table is no doubt a major flaw, for the little work that has been done indicates that it is (or was) not fundamentally distinct from any other region.

[60] It is worth considering Lévi-Strauss's (1973: 21) original passage in full: 'pour de vastes fractions de l'espèce humaine et pendant des dizaines de millénaires, cette notion [d'humanité englobant toutes les formes de l'espèce humaine] paraît être totalement absente. L'humanité cesse aux frontières de la tribu, du groupe linguistique, parfois même du village; à tel point qu'un grand nombre de populations dites primitives se désignent d'un nom qui signifie les "hommes" (ou parfois – disons-nous avec plus de discrétion – les "bons", les "excellent", les "complets"), impliquant ainsi que les autres tribus, groupes ou villages ne participent pas des vertus – ou même de la nature – humaines, mais sont tout au plus composés de "mauvais", de "méchants", de "singes de terre", ou d' "oeufs de pou"'.

Table 10.1 *The worldwide occurrence of restricted humanity.*

Ethnography and approximate location	Terms of self-reference
South America	
Tsachila – Ecuador (Ventura 2009)	*tsachila* ('true people', 'veritable us')
Huaorani – Ecuador (Rival 2002)	*huaorani* ('true human beings')
Shuar – Ecuador (Rubenstein 2002)	*shuar* ('people')
Piro – Peru (Gow 2001)	*yine* ('humans')
Ashéninka – Peru (Varese 1968)	*ashéninka, asháninka* ('men', 'people')
Matis – Brazil (Erikson 1996)	*matis, matis kimo* ('human beings', 'the only veritable humans')
Wari' – Brazil (Vilaça 2006)	*wari', iri wari'* ('human beings', 'real humans', 'true humans')
Panará – Brazil (Elizabeth Ewart, pers. comm.)	*panará* ('people')
Guajá – Brazil (Cormier 2003)	*awá* ('men', 'people')
Yanomami – Venezuela (Alejandro Reig, pers. comm.)	*yanomami* ('human beings')
North America	
Navajo – United States (Witherspoon 1997)	*dine* ('the people', 'we, the people')
Cayuse Indians – U.S. (Sturtevant 1998)	*tetawken* ('we, the people')
Tlingit – Canada (Boas 1916)	*lingit* ('people', 'humans')
Yup'ik – Alaska (Fienup-Riordan 1991)	*yup'ik* ('true people', 'genuine people')
Flathead Indians – U.S. (Waldman 2006)	*salish* ('the people')
Sioux – U.S. (DeMallie 2009)	*lakota, dakota, nakota* ('the men', 'the people')
Ojibwa – Canada (Hallowell and Brown 1992)	*anishinabek* ('the people', 'the good humans')
Iroquoian Mohawks – Canada/U.S. (Sturtevant 1978)	*onkwehonwe* ('real people')
Mi'kmaq – Canada (Fuller 2005)	*lnu* ('human beings', 'the people')
Cherokee – U.S. (Bruchac 1995)	*aniyunwiya* ('real people')
North Asia	
Mansi – Siberia (Bitkeeva *et al.* 2005)	*mansi* ('persons', 'human beings')
Chukchee – Siberia (Levin and Potapov 1964)	*lygoravetlyan* ('true, genuine men')
Yurak Samoyeds – Siberia (Golovnev and Osherenko 1999)	*nenets* ('real people')
Enets – Siberia (Levin and Potapov 1964)	*enej enet'-enche* ('the real human beings', 'the genuine men')

Table 10.1 (*cont.*)

Ethnography and approximate location	Terms of self-reference
Kets – Siberia (Levin and Potapov 1964)	*deng* ('men', 'people')
Nganasans – Siberia (Levin and Potapov 1964)	*ngano nganasan* ('genuine men')
Even – Siberia (Bitkeeva *et al.* 2005)	*inkan bey* ('real men')
Kants – Siberia (Levin and Potapov 1964)	*hantõ* ('human beings')
Nivkh – Siberia (Gruzdeva 1998)	*nivkh* ('men', 'people', 'human beings')
Ainu – Japan (Kuzuno 1994)	*ainu* ('human beings')
South Asia	
Amis – Taiwan (Mair and Bender 2011)	*pangcah* ('humans', 'people of our kind')
Atayal – Taiwan (Yoshimura 2007)	*atayal* ('genuine persons')
Kmhmu – Laos (Proschan 1997)	*kmhmu* ('humans', 'persons')
Mien/Yao – Thailand, Vietnam, Yunnan (P. Smith 2002)	*mien, mienh* ('people', 'human beings')
Iban/Dayak – Borneo (Heyward 1963)	*iban* ('persons', 'our people')
Batek – Malay Peninsula (Endicott 1979)	*batek* ('person', 'person of our group')
Chewong – Malay Peninsula (Howell 1984)	*bi he* ('people us' or simply *he*, 'us')
Ongee – Andaman Islands (Pandya 1993)	*ongee* ('the men', 'the people')
Khumi – Bangladesh (Barua 2001)	*khumi* ('the best men', the best people')
Mundas – India (Sengupta 2009)	*horoko* ('men')
Africa	
Khoisan, 'Bushmen' – Southern Africa (Barnard 1992)	*!kung* ('real people')
Khoikhoi, 'Hottentot' – SW Africa (A.B. Smith 2002)	*khoi-na* ('the real people')
Igbo – Nigeria (Ukpolo 2010)	*mmadu* ('human beings', 'the beautiful ones', 'the living ones')
Bubi – Island of Bioko, Equatorial Guinea (Aymemi 1942)	*bachoboricho* ('people of the land who are among the living')
Aka/Baka, 'Pygmies' – Central African Republic, Congo (Bahuchet 1992)	*aka* ('persons', 'people')
Bantu – Central and Southern Africa (Vansina 1990)	*bantu* ('the people', 'humans')
Uduk – Sudan, Ethiopia (James 1979)	*wathim pa* ('people', 'those of the home')

Table 10.1 *(cont.)*

Ethnography and approximate location	Terms of self-reference
Dinka – Sudan (Pocock and Macclancy 2001)	*jieng* ('humans'), *monyjang* ('the men', 'man of all men')
Anuak – Sudan (Metho 2006)	*anuak* ('the people who share, eat together')
Nuer – Sudan (Werbner 2002)	*naath* ('human beings'), *raam mi raan* ('real person', 'true human being')
Oceania	
Pitjantjatjara – Australia (Hilliard 1968)	*anangu* ('human being', 'person')
Yankunytjatjara – Australia (Eckert 2007)	*yanangu* ('human being')
Ngaatjatjarra – Australia (Glass and Hackett 2003)	*yarnangu* ('human', 'person')
Martu – Australia (Davenport, Johnson and Yuwali 2005)	*martu* ('one of us', 'person')
Koori – Tasmania, Australia (Horner 2004)	*koori* ('person', 'people')
Yolngu – Australia (Hayami, Tanabe and Tokita-Tanabe 2003)	*yolngu* ('persons', 'human beings')
Tiwi – Australia (H. Smith 2008)	*tiwi* ('people', 'we, the only people')
Yamatji – Australia (Simpson *et al.* 2001)	*yamatji* ('the men', 'the human beings')
Maori – New Zealand and throughout Polynesia (Harvey 2000)	*maori* ('the normal ones', 'the real ones', 'the genuine ones')
Aboriginal Hawaiians (Okamura 2008)	*kanaka maoli* ('the true people', 'the real people')

Eusebius Salverte (1864: 90–91), the great yet hardly acknowledged pioneer of the study of native self-designations, already recognized this in the nineteenth century. His remark on the former inhabitants of Wales is telling in this respect:

> The Welsh still retain the old name of the nation of which they originally formed a part; they call themselves Kymry, and seem to have been uniformly consistent in the practice. [. . . This name derives] from Cymrhain, family. This is undoubtedly another instance of a name given by a nation to itself; it is perhaps the oldest; we, the family, the gathering of people, the men: an isolated race can only realize its existence by some such collective idea as the above. The title *man* precedes all others.

Even though his phraseology may be somewhat dated, Salverte's principal point still holds: so-called indigenous people overwhelmingly refer to themselves as 'the men'. Contemporary anthropologists prefer to speak of 'the people' or 'the human beings', but what is crucial is that all these terms are understood in the sense of a restricted humanity; those who are included are better, more beautiful, superior, while those who are excluded are strictly speaking *not* human. However, this exclusion is not absolute, for it is never without appeal. Within animism, humanity is not only restricted but also fundamentally open. I will come back to this important point in a moment. At this stage, it suffices to underline that these notions of restricted humanity are astonishingly uniform in various respects. One of their most striking features is that they are always premised on a specific effort (Praet 2009a, 2009b). For example, to qualify as 'Chachi' or 'Human' requires, quite literally, continuous maintenance (and from here onwards I use capitalization when I refer to animistic notions of humanity).

THE INDISPENSABLE MAINTENANCE OF HUMANITY

How does this work in practice? Everything depends on what Chachi people refer to as *ura'chunu*, 'living well'. In a nutshell, this entails what could be described as 'proper marriage' and 'worthy behaviour' (Praet 2006). Above all, marriage is taken very seriously. Husband and wife are supposed to remain absolutely faithful until death separates them. Up until today this emphasis on strict monogamy is no laughing matter. Divorce, adultery and polygamy are considered among the gravest of errors imaginable. Indeed, monogamy constitutes a privileged criterion to demarcate those who are Chachi from those who are not. Blacks, Indians from the highlands and whites are considered much more easy-going when it comes to marital life, an attitude many Chachi deem inappropriate, not to say scandalous. The other crucial element is that one's conduct should be 'worthy'. In the case of men, this mainly refers to canoe crafting, hunting and fishing with spears. As for women, it primarily pertains to weaving, gardening and child-rearing. A mature man who cannot craft a canoe is often stigmatized as 'worthless' (*balejtu*). The same goes for those who are inept hunters or fishers. Anyone who fails to 'live well' in this particular way is, or at least used to be, excluded from Humanity; he or she is no longer considered Chachi. This conclusion is in line with recent findings from Amazonian anthropology.

More specifically, I refer to research that emphasizes the salience of corporeal practices in Amerindian societies. Consider the work of Vilaça and Conklin on the Wari' of western Amazonia. Wari' people refer to themselves as *iri wari'*, 'real humans' or 'true humans' (Vilaça 2006: 162). Traditionally, they concentrated on exploiting the resources of the interior forest, away from the larger rivers. Prior to contact, they had no canoes and avoided fishing in deep water. Their farming practices also oriented them away from the bigger rivers, their principal staple being maize, which grows in upland gardens (Conklin 2001: 29, Vilaça 2006: 62). Vilaça (2006: 69) emphasizes that 'maize is more than a staple crop; it is an emblem of humanity' and that 'only human beings, *wari'*, know how to cultivate it and use it as a basic ingredient of their diet'. Fatty meat and unfermented maize *chicha* are the foods that the Wari' value most highly. Giving and accepting such items, Conklin explains, expresses 'kinship, amity, and social engagement, while refusing it can imply distancing or a negation of social ties' (2001: 67, 89). I suggest that the former amounts to the necessary maintenance of Humanity, while the latter implies a negation of Humanity. Significantly, the Wari's former enemies (other Indians and the Brazilians and Bolivians, who invaded their lands in search of rubber) did not inhabit the interior forest but tended to settle along bigger rivers. Their diet was not based on maize but on such items as rice, salt and sugar. That is, they behaved in an ostentatiously non-Wari' way and were in fact conceived of as non-Human. In my view, Conklin's observation that shared bodily substance establishes kinship relations supports the more general idea that Humanity always requires a specific effort and is therefore restricted. Indeed, I would suggest that *wari'* is best translated as 'Humans', and not as 'humans' in the modern anthropological sense. In order to qualify as a Human, one has continually to conduct oneself as such and in the Wari' case this implied (and in all likelihood still implies to some extent) living in the interior forest, hunting and growing maize. This case is far from exceptional: in Amazonian anthropology it is now commonplace to say that humans are always works-in-progress that depend on the sustained maintenance of specific forms of commensality and conviviality. In other words, people are 'made' or 'fabricated' (Fausto 1999). A similar idea consistently recurs, moreover, in other ethnographic regions.

Many examples could be given, but in order to keep this chapter to a reasonable length I will concentrate on just one: that of the *ongee* – literally 'the men' or 'the people' – of the Andaman Islands in the Bay of Bengal. Pandya (1993, 2009) describes them as hunters, sea fishers and gatherers. He emphasizes that Ongees have a pronounced predilection for chewy foodstuffs; soft and liquid items such as honey, which can be consumed without mastication, are considered unsuitable for people. The importance of smell is another key element. Smell is envisaged as something that moves in and out of one's body and its control is a permanent worry. The Ongees' obstinate use of fire, clay paints and human bones has everything to do with this. Pandya (1993: 111) shows that these substances function as armour. Without them, the human body 'cracks up' as a result of the impact of the winds and the release of smell which attracts the spirits. That is why Ongees are rarely seen without fire and why their campsites are constructed around a central hearth, which protects the residents by a permanent screen of smoke. Clay paints and bone ornaments are also seen as means to restrict the movement of smell and to 'bind' it to its source, the human body. Weight is another obsession. This is linked to smell since its release always implies a change of weight. The vernacular term for 'heavy' signifies that one lives together with one's kin, that one benefits from sharing food and hence that one is doing alright. Those who live on their own, on the other hand, soon become light and hungry; they lose their smell and are easily carried away by the spirits. In my view, this indicates that Ongees do not take Humanity for granted. Those who emit smell or lose weight put their own Humanity at risk. According to Pandya (2009: 56) himself, the human subject 'is constantly haunted by the possibility of its collapse'. In order to maintain one's Humanity, a sustained effort is required. The movement of smell must be controlled at all times, while heaviness has to be promoted ceaselessly. The use of smoke, clay paints and bone ornaments perpetually reconfirms the Ongees' Humanity. Masticating food is equally important; it could be said that the essence of their Humanity is to chew. It is no coincidence that those who depend on soft and liquid food because they lack teeth (such as babies and the elderly) are envisaged as jawless spirits, that is, as non-Human.

WITHIN ANIMISM, LIFE IS A LIMITED EXTENSION OF HUMANITY

In sum, people such as the Chachi, the Wari' and the Ongee never take humanity for granted. Tentatively, I would suggest that they do not take life for granted either. Being human or being alive is not so much a safe premise as a never-ending effort which has to be kept up through such things as the continuous exhibition of worthy behaviour, the cultivation of proper food, the sharing of maize and fatty meat, the control of smell, mastication, and what one could describe in more general terms as a certain kind of commensality. Just as I have spoken of Humanity, one could speak of Life. To be precise, Life is an extension of Humanity. But contrary to the biological notion of life, where humans are just one species among millions of others, it is not a massive extension. By and large, Humans are the main component of Life. The range of living beings that are not Human is remarkably constrained. A brief look at Chachi animism will clarify this. As far as I could establish, Chachi people do not have a specific term for Life. The expression that comes closest is *chachi*, which means 'the people' as we have seen, but can also be translated as 'the living ones' or 'those who are alive'. What is very clear, however, is that they quite stringently distinguish things that are alive from things that are dead.

Pets are unequivocally alive, for example. Chachi hunters are in the habit of taking home infant animals that have been injured or orphaned during their trips in the forest. Baby peccaries, sloths, birds, monkeys and wildcats are frequently adopted into the household (see Erikson 2000, for a more elaborate account of this practice, which is widespread in Amerindian societies). They are nursed and fed with a great deal of care and affection; such companion animals are a tremendous source of pleasure for children and adults alike. Since they clearly partake in the everyday conviviality of the village they are included within the sphere of Life. This is not the case, however, for their untamed counterparts that live in the deep forest, which is conceived of as a domain of death. In a similar vein, the ocean is conceived of as a domain of death and the fauna that inhabit it and the alien seafarers that sail it belong to the dead rather than to the living. Not coincidentally, Chachi people have a notable predilection for settling towards the headwaters of the rivers, away from the estuaries – they may be prodigious canoeists but they are not (or at best, they are reluctant) seafarers. A similar limitation exists with regard to plants: the sphere of Life is roughly restricted to the herbs and fruit trees around the house, and to cultivated crops such as beans and plantains. But the status of 'being alive' is also attributed to things that modern biologists (and Moderns more generally) would not consider alive at all; this is of course a classic feature of animism. Rivers are a telling example. Chachi people do not see the whole river as a living entity, but only certain fairly well-defined parts of it, namely the stretches where the current is unobstructed and the water flows straight. Significantly, it is precisely along these stretches that Chachi have always built their settlements. Stretches where there is turbulence or a counter-current are rigorously avoided, for they are associated with death. It is not by chance that Chachi graveyards tend to be located near whirlpools. Nor is it a coincidence that all sorts of monstrous creatures, all closely related to the world of the dead, abound at such 'contrary' spots. Water Ghosts (*pillujmu*) usually bother canoeists near rapids, waterfalls are the abodes of man-eating River Dwellers (*pichulla*), and Water Dragons (*pipiñi*) are associated with the rising tides and tidal waves (Praet 2006).

This thumbnail description of the Chachi sphere of Life is admittedly sketchy, but it suffices to show that Life, just like Humanity, is fundamentally restricted. At the same time, both are also fundamentally open. The ubiquity of various practices of adoption attests to this essential openness. I have just mentioned the importance of raising captured baby animals in Chachi households: as creatures of the forest, they do not pertain to the restricted sphere of Life. But by being continually nurtured and taken care of they are literally *made* alive. The same goes for human babies: Mestizo and other non-Chachi orphans that are occasionally adopted are *made* human through various modes of commensality and conviviality. Moreover young children are not the only ones who can benefit from this possibility. Adults, even those whose physical appearance seems unmistakably different, may also be incorporated. Vilaça's (2007: 174) account of how the Wari' accepted her in their midst is emblematic in this respect:

> Among the Wari', the body of the child . . . continues to be fabricated after birth through alimentation and the exchange of body fluids with parents, siblings, and other near kin. Adopted children, for example, are considered consubstantial with their adoptive parents.. . . Food is central to the constitution of physical identity, both of the Wari' and of animal species.. . . In the first phase of my fieldwork, I repeatedly heard exclamations of the type 'She isn't Wari', she doesn't eat grubs.' When I finally swallowed some of these grubs in front of them, the news spread through the village that I had turned completely Wari'.

Just like the Chachi, the Wari' clearly insist that Humanity is earned rather than given. It is an achievement and the simple act of ingesting a grub may be the tipping point beyond which you become fully accomplished. What always recurs in such forms of animism is a Humanity envisaged as an ongoing fabrication rather than as a guaranteed status. It refers to a sustained pattern of activity or particular way of living rather than to an absolute class or category. Humanity and (by extension) Life thus possess the quality of a verb rather than that of a noun. All this suggests that Ingold's (2011: 165–76) recent argument that Koyukon names for animals describe 'goings on' is not only valid for other forms of animism elsewhere in the world too, but could even be extended to include humans. The terms of self-reference listed in Table 10.1 epitomize various composite but nevertheless relatively specific 'goings on' or occurrences. In sum, what emerges is that both Humanity and Life are simultaneously restricted and open. In many ways, they are twin notions. So far as I know, this twinned configuration of Humanity and Life has never been identified as a specific topic of research. Yet, it is far from exceptional. In Amazonian ethnography, as well as in the literature on animism more generally, there is overwhelming evidence that it is (or, at least, was) very common.

THE EXPANDED AND VERY PALPABLE SPHERE OF DEATH

I now turn to some issues that have remained in suspense so far. If Humanity and Life always require a specific effort and are always situated in a more or less clearly defined (but not static) sphere of conviviality, what about those who do not fulfil the conditions and those at the fringes of that comfort zone? If Humanity and Life are restricted, what about those who fall outside the restrictions? Focusing on how Chachi people envisage non-Chachi will elucidate a couple of things. During the early stages of my fieldwork in Esmeraldas there were persistent queries about my alimentary habits. One man asked: 'Is it true that in your country there are man-eaters?' I replied that I had never heard of such a thing, but was assured that at least in Quito, Ecuador's capital, there are whites trafficking human flesh. Such dreadful commerce was also deemed to take place closer by. In a village only one hour downstream from where I resided, there were rumours that a lodge that was being built there for American tourists was in fact a facility that would allow white man-eaters to feast on Chachi children. Previous researchers have encountered similar difficulties. A senior member of my host family recounted the arrival of John Lindskoog, a missionary from the Summer Institute of Linguistics who worked in the Cayapas river area in the 1960s, as follows:

> The first Gringo who came to this area was Juan. We were afraid and asked the local guide who was accompanying him: 'Does that Gringo eat people?' The guide replied that he did not. Later, when he already spoke

> a bit of the vernacular, Juan confirmed that he did not eat humans and insisted that he was human himself.

To be sure, this is not a recent phenomenon. The European 'conquistadors' of the sixteenth century were treated in a comparable manner. In myths, the Spaniards are explicitly opposed to 'real humans'; they were not *gente verdadero* – they were not Chachi. In the vernacular, they are designated with the expression *uyala*, just like contemporary whites. Spaniards are depicted as promiscuous, cannibalistic and warlike, entirely unlike the Chachi who pride themselves not only on their observance of strict monogamy and their skills as canoe crafters and hunters but also on their peaceful attitude. For Chachi people, these newly arrived Europeans were *not* Humans.[61] The corollary is that Spaniards and whites cannot be envisaged as 'foreigners', 'strangers' or any kind of 'other humans', for all such notions imply a shared humanity as well as a shared status as 'living being'. The difference is much more abrupt: those who are not Chachi have absolutely nothing in common with those who are – not Humanity, not even Life. Indeed, Spaniards are closely associated with death in many ways. For example, it is no coincidence that mourners at Chachi funerals temporarily transform into Spaniards, sometimes wearing colonial-style uniforms, antique swords and *morrión* helmets and occasionally murmuring archaic Spanish phrases (Praet 2005). In short, the positions of Chachi and non-Chachi are entirely incommensurable, even though they are perfectly equivalent at the same time (that is, they are equally tangible, equally visible, equally animate and so forth).

Moreover, this notable conjunction of total incommensurability and perfect equivalence appears in most if not all Chachi dealings with alterity. It does not only apply to vernacular expressions that anthropologists – misleadingly in my opinion – translate as 'foreigner'; it also applies to vernacular expressions they – equally misleadingly – translate as 'animal' or some kind of animal. Consider the following

[61] Conversely, a very similar debate emerged in Renaissance Europe, in the century after Columbus's discovery of the Americas. In a 1537 papal bulletin, *Sublimis Deus*, Pope Paul III declared that Indians are true human beings, capable of receiving the Catholic faith. In the early 1550s, the Dominican friar Bartolomé de las Casas used this argument in the famous Valladolid debate, where he disputed the widespread enslavement and exploitation of natives by the Spaniards. Such explicit statements indicate that, for many if not most sixteenth century conquistadors, the human status of American Indians was far from obvious.

illustration. In line with earlier ethnographers of the Chachi, I initially translated the term *avendyushu* as 'white-lipped peccary' or, in more general terms, as 'forest animal'. The problem with this kind of translation, however, is that it presumes a common denominator, namely 'life'. That is, I conceived of peccaries as 'other living beings'. Moreover, I projected this view onto the Chachi hunters I was working with: surely, I reasoned, the latter must think of peccaries as 'alive'? That move, I now realize, was unfortunate because *avendyushu*/peccaries are, by default, dead. To be sure, the hunters I often accompanied would agree that they are animate; usually they would even insist that they are as clever as Chachi. But that does not imply that peccaries belong to the sphere of Life. In this context, peccaries are not *just a little bit* different – they are *very* different; there is no common ground whatsoever. Properly speaking, they are not alive but dead. The Chachi hunters I worked with envisaged predator and prey, hunter and peccary, as equivalent but incommensurable positions. If one is alive, the other is necessarily dead. From this viewpoint peccaries are animate but dead. There is no term in English that captures this condition: strictly speaking, one should refer to them not as 'wildlife' but as 'wild-dead'.

This unified conception of Humanity and Life – and the expanded notion of Death that complements it – is more widespread than usually recognized; I insist that it is not the unique view of one isolated tribe in a faraway forest. Within the limits of this chapter I cannot demonstrate this at length, but in order to underscore the point I briefly return to the Wari' and Ongee ethnographies discussed earlier. Wari' people use the term *wijam* to refer to unrelated Indians as well as Brazilians and whites more generally. The term is often translated as 'foreigner' or 'enemy'. Before the intensified contacts with Brazilian mainstream society in the second half of the twentieth century the Wari' lived in an almost permanent state of war. Nowadays many of them interact frequently with whites and overtly violent confrontations have become rare. Whites are nevertheless still defined in opposition to Wari' and many suspicions remain. A perennially stressed feature is that whites live near big rivers, while the Wari' prefer to inhabit the interior forest. Their purported craving for human flesh is contrasted with the Wari's maize- and meat-based diet. In the 1990s Vilaça (2006: 496) witnessed several incidents during which people went into hiding, as they were afraid of being slaughtered by whites.

Significantly, whites are associated with *orotapan*, man-eating underwater predators dwelling in big rivers. In addition, this riverine domain is also the world of the dead. Vilaça (2006: 301) perceptively suggests that 'Whites and the dead are equivalent'. Indeed the territory of whites, whose foremost peculiarity is that they reside near big rivers, overlaps exactly with the aquatic world of the dead. In many accounts the designations *wijam* and *orotapan* are used interchangeably, which indicates that whites are envisaged neither as 'other humans' nor even as 'other living beings'. In effect, whites are simultaneously equivalent to and incommensurable with Wari'. Whites and Wari' may be equally material and equally animate but they have nevertheless no common denominator. If the latter are human, the former are not. If the latter are alive, the former are not. In other words, whites are excluded from the sphere of Humanity as well as from the sphere of Life. In a similar vein, the Wari's most preferred prey – the peccary – is excluded from the sphere of Life. Peccaries may be animate, but they are also tangible representatives of death; it is no surprise that they have their abodes underwater according to Wari' lore and that they play a central role in funerary rituals. Vilaça (2006: 129) explicitly mentions the link between prey and the dead. When peccaries and other game are scarce, she notes, it is sometimes suggested that they are hiding in the river, in the underwater realm of the dead. In sum, we see again how confusing it is to designate hunters' prey as 'wildlife'. They are really 'wild-dead'.

As Radcliffe-Brown (1922: 138) already noted, Andaman Islanders such as the Ongee have been accustomed to see all sorts of exotic seafarers visit their shores: Europeans, natives from the coasts of India, Burma and Malaya and occasionally perhaps Chinese. Such aliens, who mastered the lethal power of fire and smoke (in all likelihood guns), were associated with anger and war; they were perceived as homologous to the smell-absorbing, jawless spirits known as *tomya* (Pandya 2009: 38–39, 47). That they are conceived of as jawless indicates that they are opposed to Ongee, whose primary characteristic is their ability to chew, as we have seen. As non-masticators, alien seafarers are excluded from the sphere of Humanity. Strictly speaking they are not foreigners or 'other humans' but non-Humans. Suckling infants who have no teeth and are nurtured with soft and liquid food that does not require chewing are excluded on the same grounds – they are not (yet) Ongee or Human. In short, what demarcates restricted Humanity in this specific case is first and foremost one's ability to masticate. Moreover the association between smell-absorbing spirits, alien seafarers and the dead is quite explicit. While the living emit smell (which is why Ongees do everything they can to avoid its dispersal, as we have seen), the dead absorb it. While those who are alive are heavy, the dead are light and boneless.

To die, Pandya (1993: 53) notes, is always envisaged as becoming light and toothless. That is why Ongees deem it essential to remove the lower jawbone from the corpse at funerals (Radcliffe-Brown 1922: 313). In other words, Life is restricted to those who know how to retain their smell and their teeth and to those who are able to remain heavy. It is not coincidental that alien seafarers are identified with stones protruding from the sea, that is with 'light' stones that fail to sink (Pandya 2009: 248). Even though animate, such seafarers are excluded from the sphere of Life – they are neither 'other humans' nor even 'other living beings'. Tellingly, this quality of being animate but dead is also attributed to the favourite prey of Ongee hunters: sea turtles and an endemic species of bush pig. Pandya (1993: 92) explains that the hunt is conceived of as an eternal struggle between emitting smell and absorbing it. If the hunter is not careful – if he does not apply clay paint, wear bone ornaments and carry fire wherever he goes – sea turtles and bush pigs will absorb his smell, in which case he dies. That is, they draw him into the domain of death, of which they are tangible representatives; as absorbers, sea turtles and bush pigs are indeed on the side of the dead rather than on the side of the living. Just like alien seafarers, they are excluded from the sphere of Life – they should be understood not as wildlife but as 'wild-dead'.

CONCLUSION

At this point, I can formulate the argument in more general terms. My tentative claim is that Amerindian and other so-called animistic people do not have, or at least used not to have, any notion of 'other humans' or 'other living beings'. The very ideas of 'foreigner' and 'wildlife' were, and to some extent still are, conceptually impossible. Within animism, humans and living beings can only be subsumed under 'us' but never under 'them'. Restricted Humanity and restricted Life imply that there is no common ground whatsoever between those who are included and those who are excluded – there is neither a shared humanity (implicit in the idea of foreigner) nor a shared condition of life (implicit in the idea of wildlife). Of course, such modern anthropological and biological ideas may have been adopted by contemporary Amerindians, but they are relatively recent imports, akin to such things as the wheel, the horse or the shotgun. Before I explain this intuition in more depth, it is useful to say a few words about those who hold the peculiar belief that there are such things as 'foreigners' and 'wildlife': adherents of the western cosmology.

What is striking about this cosmology but rarely scrutinized is that it always presupposes a common denominator. For an Englishman, Frenchmen are foreigners: the French are different, but *just a little bit*. They are 'other humans', not 'non-humans'. The difference is never total – there is always a certain amount of commensurability. In a similar vein, marine biologists envisage deep-sea microbes as 'wildlife'. That implies that microbes are different, but *just a little bit*. They are 'other living beings' – not 'non-living beings'. The difference is never total. Adherents of western cosmology always presuppose a certain amount of commensurability. So for foreigners, the common denominator is humanity, for wildlife the common denominator is life. What I would like to emphasize is that this is an a priori supposition rather than a God-given truth. Even though there is a curiously uncontested consensus among modern scientists and westerners more generally, nothing actually obliges us to conceive the world in terms of commensurability.

In my understanding, so-called animistic people de facto reject (or used to reject) this assumption of a common denominator. If that is correct, then imputing the idea of 'other humans' or, more generally, that of 'other living beings' to people such as the Chachi, the Wari' or the Ongee is fundamentally flawed. At best, such ideas are colonial or postcolonial coinages whose European roots remain unacknowledged. Those who fail to see this unwittingly end up describing the people they study as minimal anthropologists and as minimal biologists, in the sense that they project their own innermost convictions (the belief in 'other humans' and the belief in 'other living beings', respectively) on their subjects of inquiry. In this way, my account of restricted Humanity and restricted Life has put the inclusive humanity and the inclusive life of the western cosmology into perspective.

At first glance, this emphasis on restriction may seem like a retrogressive step. After all, the overall trend in current anthropology has been to open up the discipline to previously marginalized animals, vegetables, fungi and whatnots. The inclination of recent projects such as Kohn's (2007) 'anthropology of life' or the 'multispecies ethnography' advocated by Kirksey and Helmreich (2010) is to include rather than to exclude. What I have proposed in this chapter does not necessarily contradict this tendency, however. For one thing, exclusion from Humanity or Life is never without appeal. The fact that they are eminently conditional implies that they are not only restricted but also fundamentally open. Those who put in the appropriate effort and receive adequate support eventually become human or alive. Both Humanity and Life can be *achieved* by those initially excluded: injured baby forest animals, alien foundlings and young orphans, and even the occasional well-disposed ethnographer. Clearly, there is no absolute barrier. However, this orientation toward achievement does not mean that anything goes. The room for manoeuvre is usually quite limited. Reckless Chachi experimenters who deviate too much from core ideals such as strict monogamy, good craftsmanship and skilful spear fishing risk being excluded (even though there are various procedures to keep them in check; Praet 2009a, 2009b). The inherent conditionality that characterizes animistic notions of Humanity and Life entails an expanded notion of Death which comprehends all those whom western observers traditionally designate as 'foreigner' and 'wildlife'. Ethnographers may wince at the prospect of having to deal with such an unexpectedly rich and palpably present domain of Death. But once one realizes that what is dead is not necessarily rotten, repugnant, bad or inanimate, and once one grasps that this specific set of associations is a parochial (if remarkably widespread) dogma rather than a universal truth, the step is easy to take. In a text to which I have already referred, Landecker (2003) identifies the strange 'lack of death' that characterized twentieth century cytology. The present account of Chachi and other forms of animism reveals a similarly strange lack of death at the macroscopic level, in contemporary anthropology and biology.

ACKNOWLEDGEMENTS

I am very grateful to Gisli Palsson and Tim Ingold for giving me the opportunity to present the paper from which this chapter originates on their panel at the 2010 EASA conference in Maynooth, Ireland. I thank the panel discussant, Deborah Heath, for her stimulating observations. And warm thanks to my fellow panellists for their inspiring comments and good company. I have greatly benefited from the generous support of the Wenner-Gren Foundation (Hunt Fellowship), from the Rachel Carson Centre in Munich (Carson Fellowship), as well as from an internal research grant from my own department at Roehampton University.

HAYDER AL-MOHAMMAD

11

Ravelling/unravelling: being-in-the-world and falling-out-of-the-world

Reconsidering Wittgenstein's 'forms of life' (2007), Cavell arrives at the notion of 'whirls of organisms' (1976: 52) as a more faithful way of rendering the constitution of persons as grounded in what they *do* and not in what they *are*. Where philosophers have tended to obsess about the *forms* in 'forms of life', Cavell is more interested in *life* and its contingent and haphazard, though necessary movements and interlacements through the lives of others. Yet Cavell's 'whirl of organisms' does not carry any sense of binding tie or relation; it rather brings to mind William Blake's capricious 'howling storm'. In *The Palm at the End of the Mind*, however, Jackson picks out a greater sense of mooring in the world in the overlapping of lives, when he writes of 'a sense of being grounded in something I can only call "the real" that connects my life to the life of the earth itself' (2009: 1). This most intimate connection between self and world, one which we inhabit and exist by, seems estranged from comprehension for Jackson. Wittgenstein tracks this estrangement to an existential closeness which obfuscates what is most fundamental to us: 'The aspects of things that are most important for us are hidden because of their simplicity and everydayness. (One is unable to notice something – because it is always before one's eyes)' (2007: I, 129).

Of what order, if any, is the connection between person and world? I shall investigate this connection by running two parallel narratives: one theoretical, the other ethnographic and eventful. The theoretical narrative is focused on Heidegger's account of 'being-in-the-world'. Heidegger's position on the relation of persons and world is much more sophisticated and nuanced than many within the

Biosocial Becomings: Integrating Social and Biological Anthropology, eds T. Ingold and G. Palsson. Published by Cambridge University Press.

humanities and social sciences have recognized and in part this chapter is an attempt to complicate our understandings of Heidegger and phenomenology. But this aim is secondary to what I am truly interested in, which is how we can think of and account for the connectivity, and also the disconnectivity, of being and world. In thinking through this problem I shall turn for help to Ingold's trope of the 'meshwork' as a way of conveying a more processual and dialectical account of inhabiting the world. The ethnographic events I recount below are intended to disrupt the somewhat cosy theoretical narratives of human being as necessarily or fundamentally at-home-in-the-world.

I begin with a definitional outline of 'being-in-the-world' as it has been conceived by some anthropologists and by Heidegger scholars. My stress is on the necessary relationship of 'being' and 'world' and how best to think of that relationship in a way that allows us to discuss social being without marginalizing or diminishing the experiences of others. I move on to a brief ethnographic vignette of Adel, a guard at the hotel in Basra, Iraq, at which I regularly stay. I do not use Adel's case to exemplify any theoretical concepts; instead I want to look at how, through one person's experiences, we can better articulate those intimate and rarefied categories (such as life, existence and being) which constantly resist determination. In the next section I highlight some of the ways in which Heidegger's thought overlaps with that of Aristotle, in order to show how everyday practical immersion in the social world comprises both the context of our experiences and also its metaphysical grounds. By not splitting experience from a distant, otherworldly metaphysics, we are able to take much more seriously the daily pragmatics of human existence. In subsequent sections I show how precarious our fix in the world can be and how easy it is to fall out of the world. In this light, I go on to reconsider being-in-the-world as a much more fraught affair than is commonly supposed, a position which Heidegger holds too in *Being and Time*, as I hope to elucidate. If, as Ingold maintains, lives are lines entangled and enmeshed in a meshwork of other lines, then we need to introduce a greater sense of instability within the meshwork by drawing attention to both the contingency of our being 'enmeshed' and the simultaneous and dialectical movements of ravelling and unravelling.

BEING-IN-THE-WORLD

One of the difficulties confronted by the social sciences in recent decades has lain in attempting to overcome what Strathern calls the 'mathematic' which 'sees the world as inherently divided into units' – that is, individuals – the aggregate of which is 'society' (Strathern 1996: 49). Many have found Heidegger's phrase, though not necessarily his account, of 'being-in-the-world' a fertile and evocative way of thinking of the necessary entwinement of world and human being without falling into mechanistic modes of relating parts to wholes. Richardson claims for Heidegger's phrase, for instance, that it reveals the symbiosis and interdependence of human being and world:

> Being-in-the-world means that for us to be we must have a world to be in. We cannot otherwise exist. Yet 'world' is not an external thing, existing apart from our actions and awaiting our entrance; but it is dependent upon our being in. Through our actions, our interactions, we bring about the world in which we then are; we create so that we may be, in our own creations. (1982: 421)

Weiss (1996) presents one of the more nuanced accounts of such a position by drawing attention to how ordinary and mundane experiences of the everyday construct a 'lived world' which grounds certain modalities of social being. The ordinary world then becomes both a condition and a product of our everyday practical engagement.

Only distinct entities, however, can 'mutually constitute' one another. For example, it would be nonsensical for me to talk of myself 'mutually' doing anything with myself. Likewise with 'constitutive': in what meaningful sense could one talk of *one* constituting *one's* self? Thus, to avoid falling into grammatical inconsistency (Wittgenstein 2007: I, 371–373), one would have minimally to posit two separate entities in a 'mutually constitutive' relationship. Though the relationship may be metaphysically prior to the existence of each entity thus related (be it 'being' or 'place'), we are still dealing with two entities with a certain bearing on one another. Being, however, does not constitute world and nor does world constitute being in Heidegger's formal description of 'being-in-the-world'. If being constituted world then Heidegger would revert to Idealism (Guignon 1983: 20–28). Alternatively, if world constituted being then Heidegger would be some sort of material determinist and would have fallen into error by hypostatizing being and assuming a super-transcendent entity at the origin of our existence (like Aristotle's 'unmoved-mover'), a position he calls 'onto-theology'.[62] Rather than thinking of entities of

[62] By this Heidegger means two things: firstly, that 'being' is a property or predicate of all beings; and second, that this ill-conceived property (which western

certain sorts as constituting other entities (such as gods and humans), the equi-primordiality of 'being' and 'world', captured by Heidegger in 'being-in-the-world', attempts to surmount the rationalistic models of modernist philosophy.[63]

The classical and most important model that Heidegger is trying to overtake is the Cartesian picture of two distinct worlds (*res cogitans* and *res extensa*) which somehow coincide within the human form. It is all too easy for a sceptic to claim that there is doubt as to the existence of the external world or as to the veracity of any synthetic claims we make about the world once the two-world account is posited.[64] Heidegger has therefore to be careful to avoid positing two distinct entities or substances such as 'mind' and 'space' or, in Robinson's reading (cited above) amongst others, 'world' and 'being'. His argument is that human being is always already involved and immersed in the world it inhabits. Though he falls back on the Kantian language of the 'condition of the possibility of experience', he does not claim that world and being constitute each other, for to do so would entail falling into the Kantian 'transcendental analytic' of locating the 'conditions of possibility'. Dreyfus clarifies the matter as follows:

> Heidegger calls his investigation an *existential* as opposed to a *transcendental* analytic. According to Heidegger's existential analytic, Dasein must exist factically – that is, Dasein (unlike the transcendental ego) is necessarily involved in, and dependent upon, the world it opens up, and can never get clear about the world it is in. Only thus is it the condition of anything showing up for us as anything. (1991: 344)

metaphysics claims to represent when it represents beings) fails to recognize the way being conditions its own representation, therefore eluding representation itself. Metaphysics represents beings in a certain light, claims Heidegger, without being able to see the light that allows beings to *be* in the first place (Heidegger 1998: 279–282).

[63] In Heidegger's own terms: 'Being-in is not a "property" which *Dasein* sometimes has and sometimes does not have, and *without* which it could just *be* just as well as it could be with it. It is not the case that man "is" and then has, by way of an extra, a relationship-of-Being towards the "world" – a world with which he provides himself occasionally' (Heidegger 1967: 84; original emphasis).

[64] To be brief, the sceptic could take several stances. The two most prominent are: (1) If I only had access to the mental representations of the world in my mind, what guarantee do I have that the outside world exists? (2) Even if the external world does exist and objects in the world are the causes of my perceptions, do my perceptions of these objects necessarily coincide with the objects themselves? One stance picks on ontological scepticism: maybe the world does not exist? The other picks on epistemic scepticism: how do I know my experience corresponds to the objects I am experiencing? (Landesman and Meeks 2003).

Contrary to modernist metaphysics, claims Heidegger, the subject-object relation (presence-at-hand[65]) is less original than the self-transcendence of being-in-the-world by which human being opens itself to the world before all knowledge and being. Even before there are entities and things which show up in experience, human being is already open to the world since 'knowing has the phenomenal character of a Being which is in and towards the world' (1967: 87). Knowing is grounded beforehand in a being-already-alongside-the-world, but this cannot be a starting point that begins with entities which are present-at-hand because 'Being-in-the-world, as concern, is fascinated by the world with which it is concerned' (1967: 88).

What is metaphysically prior for Heidegger, then, is not that human beings have minds or souls, or are of a precious otherworldly, ghostly substance. Human being, ontologically, lies in its openness and involvement to and in the world – in small part this explains why Heidegger uses the term '*Dasein*', or 'being-there', to denote human being: human being is only its being-*there*, its factical existence and skilled immersion in the practicality of daily life.

Ingold carries much of Heidegger's concerns in the early chapters of *Being and Time* into his own work, his 'ontology of dwelling' being a case in point, where he takes the human condition to be ...

> that of a being immersed from the start, like other creatures, in an active, practical and perceptual engagement with constituents of the dwelt-in world ... apprehending the world is not a matter of construction but of engagement, not of building but of dwelling, not of making a view *of* the world but of taking up a view *in* it. (2000: 42)

Ingold takes up the problem of how anthropologists can think and talk of the fundamental grounds of human being without falling for the type of classical metaphysics that proposes either metaphysical dualism or monism of a materialist or spiritualist kind, and very little else in between.

Whereas anthropologists could treat human praxis as a certain modality of social life – which in the last decades has proved popular with practice theorists of socio-political organization and control, with descriptions of how 'techniques of the body' (Mauss 1973) come to

[65] *Vorhanden*, or presence-at-hand, is Heidegger's term for encountering entities as discrete objects. Heidegger calls the entities revealed through our involvement with them *zuhanden*, or 'ready-to-hand', in opposition to *vorhanden* entities. Ready-to-hand entities are thus characterized by the specific use that we make of them (1967: 98–121).

form behavioural, perceptual and intellectual dispositions – in Heidegger's phenomenology human praxis is not the sum of a set of facts or contexts that human being inhabits, such as could be described and accounted for by the ethnographer. Nor is it enough to regard social life as a deeply important but nevertheless contingent aspect of human existence. Rather, human being-in-the-world, in Heidegger's account, comprises the very ground of that existence (Jackson 1983: 339–40, Csordas 1990, 1993: 8).

BASRA

From my initial visit to Basra in summer 2005, and throughout the period of fieldwork which ended in autumn 2009, I have always stayed in the same hotel located near the centre of the city, *al-ashaar*. One of the first people I met in the hotel was Adel, the security guard. Whenever he would finally appear for his 12-hour shift which began at 7 in the evening, he would rush to the hotel's kitchen, pour himself a large mug of tea and stand outside chatting to whoever was prepared to have a chat with him, or run off somewhere on an errand he claimed was urgent, or disappear to sleep on some cardboard next to the hotel generator where he was well hidden from view. He had to be back to his station in the front of the hotel before 11 in the evening, however, as the Sheikh[66] who was running the hotel on behalf of the owner would return at that time to check the takings of the restaurant and the hotel for the day. The Sheikh would ask to see all the hotel's employees and would consider it a personal offence if they were not fulfilling the roles for which they were being paid. Those of us in the hotel, workers or guests, would either rush to find Adel to wake him or shout down the road in the hope he might hear us and return before the Sheikh arrived.[67]

[66] Though the Arabic word 'Sheikh' has many senses, in this chapter it will only be used to refer to the head of a tribe (*'ashira*).

[67] Due to the violence in Basra – sectarian or otherwise – and the instability of the region, from 2005 until the end of 2008 there were hardly any guests at the hotel, particularly as staying in a hotel in Basra suggested that one was not a local and possibly had money which increased the risk of being kidnapped. The few guests there were tended to be involved in long-term business projects. The guests would return to the hotel before late afternoon to avoid the evening troubles. Thus, hotel guests and staff spent much time together and we formed strong relationships.

Adel was not particularly liked by the other hotel workers. He was not related to anyone in the hotel and his tribe was small and considered to deal in petty crime and little else. In the hotel Adel was the butt of many jokes – we all were, but Adel more so than the rest. During the peak of the violence in Basra, in 2006 and 2007, the militias would do battle with each other in front of and behind the hotel, and mortars would fly overhead, directed somewhere in the region of the Palace where the British army was resident, though often landing anywhere but there. Shootings, kidnappings, rape and the sight of dead bodies on the streets were not unusual during that period. Such circumstances made the figure of Adel, or rather the position he was in of having to be the first line of protection against any attack on the hotel or its occupants, seem utterly ludicrous. Adel was, at best, a security guard in name only.

At night, after the Sheikh had left, Adel would sleep on the hotel doorsteps, with an inadequate fence between him and the street. He was bitten by rats, bugs and even dogs; stray bullets glowing red and yellow in the dark night would stream across the sky as he tried to get some sleep. He would wake up at 5 or 6 in the morning, breakfast quickly and run to the open market. From 7 in the morning until 4 in the afternoon he could be seen selling remote controls, or batteries, or whatever was to hand. Under the extreme heat of the summer sun, amongst tens of other sellers, Adel would shout and scream what he was selling at passers-by. He was not the only one shouting and competing for the attention of buyers; bigger stalls had loudspeakers which drowned out his voice and some had a cassette-player on a loop playing out through loudspeakers the product being sold and the price. By the time he would arrive for his shift at the hotel in the evening his throat was sore, his voice coarse, his skin scorched by the heat, and his face dazed and tired. Many days he sold nothing, and with six daughters (two at the University of Basra) and three sons – only one working, the other two were at either a technical college or Basra University – his dejection never seemed to reach the level of panic; instead he was resigned to play this all out again the next day. The maths were simple: his monthly income as a security guard was approximately $160 and six days a week in the market rarely brought in more than $120 in the month. His rent was a flat $200 and from somewhere he had to find enough to feed and support his family.

Adel's position was always precarious. He could not fulfil the role he was being paid for and his attitude to his job was lackadaisical at best. On rare occasions Adel could be lucky. Some years ago he was resting on the reception bar with a large mug of tea telling one of his tall tales. Everybody was laughing. I walked in amidst the fuss and Adel warned me that in a few months he would be giving up his job and would stay in the room opposite mine in the hotel. 'Did you marry a rich woman?' I asked. 'Better. I was sleeping and needed to urinate. It was about 5 am. I get up and go behind the palm-tree in front of the hotel door and am about to unzip my trousers when Abu-Mustafa [the owner of the hotel] spots me, thinking I am doing my morning checks. "We need someone like you who is prepared to stay awake all night", he said to me. I'll be managing his businesses in Basra in a few weeks . . . you had all better show me some respect . . .' The hotel staff and I were all laughing; we knew Abu-Mustafa was desperate to be rid of him. He thought Adel to be unclean, stupid (*hmaar*, 'a donkey') and an embarrassment to the hotel, but this bit of luck would keep Abu-Mustafa quiet on Adel for a while.

PRACTICAL REASON AND BEING-AT-HOME-IN-THE-WORLD

Adel's ability to maintain his job at the hotel even though several of the employees wanted to be rid of him, so that they could bring one of their relatives to fill his post, or his ability to stand amongst hundreds of other sellers on the street around the grocer's *souq* in *al-ashar* and to shout and scream his way to being noticed by passers-by in the hope they would buy whatever he had to sell that day, is only a small part of the relentless work that Adel, and his family, put in to try to sustain themselves. A visual image of one of the *souqs* in the centre of Basra would not capture the heat, the smell and the violent sounds of hundreds of megaphones announcing, on repeat, the products that each vendor had to sell, and their prices. In this great sonic battle Adel, one voice amongst a cacophony of mechanical sounds, shouts what he is selling for 7 or 8 hours every day, 6 days a week. 'The sound from a person's voice is different. My throat and head hurt every day but I tried the megaphone and I did not sell anything.'

In book six of the *Nichomachean Ethics* Aristotle introduces his distinction between the two intellectual virtues: *sophia* and *phronesis*. Whereas *sophia* is intellect dedicated to the scientific and universal, *phronesis* (practical wisdom) . . .

> is concerned with ultimate particulars, since the actions to be performed are ultimate particulars. This means that it is at the opposite pole from

> intelligence. For intelligence grasps limiting terms and definitions that cannot be attained by reasoning, while practical wisdom has as its object the ultimate particular fact, of which there is perception but no scientific knowledge. (2009: 1142a 23–29)

Adel's understanding of the *souq* is not based on analysis of surveys conducted with a certain sample of potential customers, nor do focus groups provide his insight into how best to make a living in the market. 'We didn't have anything. After the invasion [March 2003] there was no work and money. I took what we had in our house to sell in "*souq al-haraj*". Our things ran out and I saw others selling small things like cheap Chinese torches. I bought a few and stood there in the *souq*. Slowly I became used to the people, the fighting, the shouting ... Now I'm the Sheikh of the *souq*.' Experience and encounters shape, implicitly, how Adel comports himself in order to function in a market in which he has constantly to negotiate and renegotiate what buyers want and how to withstand the onslaught of sellers competing for precious dollars. 'The *souq* is never stable. We have months when there is money and just like now, it has been dead since the *arb'eeniya*.'[68]

The daily pragmatics of Adel's or any person's life are neither anterior to nor effects of deeper metaphysical conditioning. For Heidegger the import of *phronesis* lies not merely in its emphasis on the practical intelligence of 'being-in-the-world', as against disengaged theoretical apprehension. *Phronesis* constitutes a mode of insight into human-being's concrete situation: both our practical situation and, more fundamentally, our existential situation. It is a form of pre-conceptual practical self-understanding. Volpi (1994: 205–209, 1996: 54–58) claims that Heidegger's *Dasein* is *phronesis* ontologized. This ontologization of *phronesis* has two aspects: first, the means by which humans engage with the world practically also ground human being; second, human being in its concernful encounters in the world is attempting to get clear of itself and its world within the horizon of its concern: thus *phronesis* is a kind of everyday intelligibility that nevertheless must come to a position on the *being* of the everyday.

One of the means by which we can come to recognize the ontological significance of practical wisdom is through our pre-reflexive use of everyday objects. To turn to Heidegger's famous example, when I am hammering the hammer itself recedes as an object; I do not think of the

[68] The *arb'eeniya* or 'fortieth' is the 40-days remembrance of *ashura* in which the Prophet Mohammad's grandchildren and his followers were killed in the battle of Karbala.

hammer *qua* hammer but focus on the nail I am hammering. Hence, the hammer in my hammering is not an object with a certain set of properties which the mind recognizes and manipulates according to some intention. The most basic relation humans have with the world around them, claims Heidegger, is one not of disengaged contemplation but of active concernful immersion.

This practical wisdom (*phronesis*) which resides in the everyday is an engagement that ties directly to a sense of homeliness: being-at-home-in-the-world is something which persons actively work and strive for (Jackson 1995: 89–94); it is at the very heart of the project of making a habitable place in which to dwell. It is in this sense, perhaps, that we should understand the term 'care': my actions in the world are ultimately bound up *in* and directed *by* concern. This is the intentional structure of experience[69] which is always directed outside of, and always transcends, human being towards the world (Lévinas 1990: 33–53). That is, praxis is always directed, and directed in terms of its concern, of its being. As being (*Dasein*) for Heidegger is always being-with (*Mitsein*) (Heidegger 1967 [1927]: 149–169), solicitude and our sense of *homeliness* must be social and tied into the world of others (Al-Mohammad 2010).

UNRAVELLING AND DISENTANGLING

On February 2009 I was sat outside the hotel's restaurant in Basra. I was waiting for Adel to appear for his evening shift to joke and chat with him. 'Your friend is not coming back'. Abu-Ali said this with a smile. 'He was told to stay at home.' 'What happened?' 'I spotted that Adel had

[69] The first formalization of 'intentionality' as the structure of experience is to be found in Brentano's famous passage in *Psychology From an Empirical Standpoint*:

> Every mental phenomenon is characterized by what the Scholastics of the Middle Ages called the intentional (or mental) inexistence of an object, and what we might call, though not wholly unambiguously, reference to a content, direction toward an object (which is not to be understood here as meaning a thing), or immanent objectivity. Every mental phenomenon includes something as object within itself, although they do not all do so in the same way. In presentation something is presented, in judgement something is affirmed or denied, in love loved, in hate hated, in desire desired and so on (Brentano 1995: 68).

Not until Heidegger, however, does 'intentionality' move out from being a problem of consciousness to one of human being's concernful action in the world.

signed in yesterday when he had not shown up for work and told the owner. He said that shit was never to come near the hotel again.' 'When did this happen?' 'Just now. I was looking at the book and saw that he had signed for yesterday, today and tomorrow ... I took it to Abu-Mustafa and that was that.' That evening I saw the hotel owner with whom I would normally have breakfast on the rare occasions when he would be in Basra. I asked if there was any chance of his reinstating Adel. 'None at all.'

Adel had a stall in one of several markets in Basra, he would rarely keep to one location. By a stall I mean nothing more than a blanket on the ground with whatever bits he was selling. The next day I looked for him in the various markets in the centre of Basra and could not find him. One of the men Adel was constantly fighting with in *souq-al-haraj* was packing up for the day. I asked of Adel's whereabouts. He swore at Adel, called him a thief. None of this was helpful. After a week or so had passed even the workers in the hotel were asking of his whereabouts. What struck all of us in our conversations about Adel's plight was that even in his poverty he managed to have three of his children accepted into Basra University and doing well, a feat that none of the hotel workers or even any of the richer guests at the hotel could achieve.

Adel lived in an area of Basra called *khamsa meel*, which literally translates as '5 miles', as it is located roughly 5 miles from the city centre. One of the poorest districts in Basra, it is not a place that outsiders would usually enter. One day I had a funeral to go to in the area and with a few of the hotel workers, we decided it was a good opportunity to find out Adel's whereabouts. Within half an hour of searching a few people had mentioned an area where we could find his family. 'We haven't seen him for a while. He's nowhere. How am I meant to raise his children? What disaster has God thrown on my head?' I walked away in the midst of what I assumed was Adel's wife's shouting. Later in the car it was forgotten about, 'if you worry about this family then there are others, and others still ... we can't even look after ourselves ...'

BEING-*IN* AND FALLING-*OUT*

How is someone able to fall out of the world, as Adel did? If human *being* is being-in-the-world, and being at home in the world is not a contingent fact about human being but ontologically primordial, and if, as Ingold has so evocatively argued, lives are to be thought of as so many lines entangled in a mesh, then how is it possible for a life to slip out of the enmeshment of our necessarily worldly being? Does phenomenology fall flat as soon as we encounter those none too rare moments when persons, like Kafka's 'K.' in *The Trial*, go out the back door, so to speak, *wie ein Hund*.

Zizek (2001: 106–107) notes with some irony how Anglo-American readings of *Being and Time* tend to stop with the first division of Heidegger's text. There is plenty in this division to suggest that our conventional picture of how Heidegger refutes representational theories of mind, external world scepticism, substance dualism, and so on, is partial and incomplete at best, at least in terms of what he was really trying to achieve. With the second division of the text, however, this picture is no longer even tenable. Before I proceed to discuss what I believe are the deficiencies in our usual account of 'being-in-the-world', it is important to bear in mind that the problem is not merely exegetical. I am not out to show that the text has been misread. What is at stake, rather, are the very fundamentals of how we think of, and attempt to account for, human sociality.

If human existence is being-in-the-world, how is it possible for human beings to be disembedded from the world which primordially we must be *in*? In our everyday state, Heidegger suggests, persons do indeed enjoy the sensation of being-at-home, such that they are able to take on their social positions: 'This characteristic of being-in was then made more concretely visible through the everyday publicness of the "they" which brings tranquillized self-assurance – "Being-at-home" with all its obviousness – into the average everydayness of *Dasein*' (Heidegger 1967: 233). However in the experience of 'anxiety', in which persons feel a discontinuity with the world of 'tranquillized self-assurance', 'Being-in enters the existential "mode" of the *not-at-home*' (1967: 233). Thus, when human beings experience an uncanny discontinuity between themselves and the world, their relationship to the worldly place shifts from a feeling of being-at-home (connectedness) to one of not-being-at-home (disconnectedness). However, if such is Heidegger's position, then we may be faced with a contradiction rather than an ambiguity regarding whether humans are at-home in the world or not-at-home in the world. Is the sense of being-at-home to be regarded as more 'real' than that of not-being-at-home, or vice versa? Or are they to be thought of as equivalent in existential weight? Heidegger's answer is unambiguous: 'Tranquillized, familiar being-in-the-world is a mode of the uncanniness of *Dasein*, not the other way around. *Not being-at-home must be conceived existentially and ontologically as the more primordial phenomenon*' (Heidegger 1967: 234, original emphasis).

This passage from *Being and Time* should make clear, initially, that for Heidegger being-in-the-world is a much more fraught existential structure than conventional readings imply. World and being, as has been repeated many times, are not entities, let alone separate and separable, yet human-being's most fundamental existential grounds in the Heideggerian account are unsettled and unsettling. There is a long tradition within the discipline of anthropology of counterpoising continuity with discontinuity and of accounting for the dialectical movements within the social between the poles of structure and anti-structure (Leach 1970, Turner 1995). With Adel's disappearance and Heidegger's phenomenology of unhomeliness, however, we can begin to think of the discontinuous as inherent in the continuous, such that each pole is coincident with and constitutive of the other.

The fact that Adel was even able to fall out of the social world means that we must qualify the anthropologico-phenomenological prioritization of skill over propositional knowledge, knowing-how over knowing-that, homeliness over homelessness. In the phenomenology of skill and embodied practices, for example, anthropologists have tried to think past the idea of knowledge acquisition as an introjection of propositional content, from a 'master' or a teacher, into the mind of the novice. Marchand (2001, 2003), in his study of minaret building and apprenticeship in Sanaa, notes that apprenticeship proceeds without explicit verbal orders and any clearly defined systematic programme, through an emphasis on inculcating obedience and disciplined comportment, devoid of any explicit understanding of the tasks being exercised or their relation to the overall project. What is missing from such an account of the embodied, pre-linguistic nature of knowledge, skill acquisition and understanding in apprenticeship, however, are the failures, breakdowns and inconsistencies in our practical and intellectual edifice: embodiment must be tied to an account of disembodiment. As persons we do not progress so fluently from apprenticeship to mastery, or in the terms of Lave and Wenger, from 'legitimate peripheral participation' to 'full participation' (Lave and Wenger 1991: 37, see also Götsch, Chapter 7, this volume).

In an attempt to grasp the dialectical and processual relation between person and world, Irving (2004) examines the breakdown that occurs when the 'habitual unity of mind, body and world . . . forged through the skills and practices of "everyday life"', disperses and fractures. Focusing his work on HIV/AIDS sufferers in Uganda, he notes that 'the sense of continuity becomes continually displaced as they have to repeatedly *re-inhabit* their bodies during and after the disruption of illness' (Irving 2004: 322, original emphasis). Sufferers have to articulate modes of learning situated within a body that cannot support the practices and assumptions which underwrote their being-in-the-world prior to their illness. They become, so to speak, *apprentices* to their illness, testing and reconfiguring the relations of their body to familiar surroundings and practices that, through the affliction, have become strange to them. By walking with HIV/AIDS sufferers, Irving notes the discrepancy between before and after becoming ill and sees this as a physical communication of the body's deterioration which disarticulates the stabilizing and silent narratives of the previously experienced unity of habitude and practice. Contours and surroundings that appear constant from the perspective of a healthy body betray an inherent dynamism for bodies that are being 're-inhabited'. This inherent dynamism can help us understand the relation between being-in-the-world and the not-at-home.

RAVELLING/UNRAVELLING

The Oxford English Dictionary reports the following senses of the word 'ravel': 'To entangle; to confuse, perplex; to render incoherent or muddled'. Heidegger claims of western metaphysics that it has forgotten the question of 'being', and perhaps there is a sense in which being-in-the-world, in its inevitable entanglement and enmeshment in the lives and beings of others, can only be 'confused' and 'perplexed'. Being-there, ravelling in the lines of other lives, occludes the process of its own entanglement. Whereas the products of beings are manifest, the very producing and productivity of being is obscured (Heidegger 1985, Schelling 2004).

In the dialectic of embodiment and disembodiment, and of being at-home-in-the-world and not-at-home-in-the-world, we can discern a much more perilous and dynamic understanding of human being: as humans are ravelled into the world of others there is simultaneously a process of unravelling which is not so much its counterpoint as part of being's very movement, its productivity. Ingold's 'enmeshment' is a somewhat cosy affair because it takes as its problem the coincidence of human-being *in* and *with* the world. By regarding human being, instead, more in terms of simultaneous movements of ravelling and unravelling, we can begin to situate finitude, disease, sickness, anomie, loneliness, despair, rage and existential trepidation within the same life movements that might be considered to be their opposites. Falling-out can be as much about trying to inhabit the world as skilfully coping.

When Adel went missing and did not return, what was highlighted was the contingent and fragile purchase his life-line had on the lines of others. There are several aspects to the entangling of lives which bear thinking about: firstly, there is the almost a priori entanglement which Heidegger captures in claiming being (*Dasein*) to be necessarily being-with (*Mitsein*); then there is what I am tempted to call the 'labour of being': the physical, emotional, intellectual, practical work that goes into trying to sustain not only our existence, but the kinds of existence we want to lead. Jackson uses the phrase 'struggle for being' to denote something similar to what I am trying to capture here. In his own words:

> Though human existence is relational – a mode of being-in-the-world – it is continually at risk. This implies not only that our being is conditional on our interactions ... with the world in which we live, but that we are involved in a constant struggle to sustain and augment our being in relation to the being of others, as well as the nonbeing of the physical and material world, and the ultimate extinction of being that is death. (2004: xiv)

Where I would go further than Jackson is in my claim that 'struggle' is not just what being *does*, but what it *is*. To reconfigure his terminology slightly, I am more interested in 'the struggle *of* being' (struggle as an ontological state) than in 'the struggle *for* being' (struggle as a predicate of being).

This aspect of the entangling of human being, its struggling productivity, in which persons actively work to entangle themselves further in the meshwork, or to disentangle themselves from it – or even try to sustain the position they believe they have already – is in my view the properly perilous movement of human being. The peril resides in persons' trying to project their lives forward, and finding that they are confronted with the openness and contingency of a world that always poses the possibility of a project's failure, or less dramatically, of its being changed and not affording the fulfilment of what was hoped for (Bourdieu 2000: 208–10). Critchley captures wonderfully the recoiling movement of projecting one's life-project forward in his analysis of Heidegger's 'throwness'.[70] Human-being:

[70] On the meaning of projection, Heidegger has this to say: 'Projecting has nothing to do with comporting oneself towards a plan that has been thought out, and in

> it is at once thrown and the projection or throwing-off of thrownness . . . Dasein is always sucked into the turbulence of its own projection. I imagine it is a little like driving a car without a windscreen: the faster you drive, the greater the resistance. Dasein is the name of a recoiling movement that unfolds only to fold back on itself. Its existentiality, its projective being-ahead-of-itself, is determined through and through by facticity, it is always already thrown in a world. (Critchley 2002: 155)

Our being-in-the-world holds no guarantee of the success or otherwise of our practical engagements. In many respects, Adel was in a much better position financially and socially than many others in Basra but this does not safeguard, existentially, very much. The labour of being, of making a habitable space for one's self and family, of attempting to get a better foothold in the meshwork of social life, unravels its entwinements in order to get a better or different purchase on the world.

In the movements of entanglement are opened the possibilities of disentanglement. Agamben presents a similar account of human being's being-there which must be that of a being which can never be *fully* there:

> It is here that one must see the root of *ausweichende Abkehr*, 'evasive turning-away', and of the impropriety constitutive of Dasein. It is because it must be its guise that Dasein remains disguised – hidden away in what opens it, concealed in what exposes it, and darkened by its own light. Such is the factical dimension of this 'lighting' (*Lichtung*), which is truly something like a *lucus a non lucendo*. (Agamben 1999: 109)

There is always something that cannot be accounted for, that cannot be fully inserted into the totality of being-in-the-world. Rather than this being taken as a grand metaphysical statement I hope that its implications for thinking about the connection between person and world are now clear. As persons we are never totally absorbed into the world around us, only thus can we account for persons falling out of the world. In his famous essay on 'The sorcerer and his magic', Lévi-Strauss shows how a shaman, after being exposed as a fraud by the mercurial Quesalid, disappears with his family in shame: 'He returned a year later, both he and his daughter had gone mad. Three years later

accordance with which Dasein arranges its Being. On the contrary, any Dasein has, as Dasein, already projected itself; and as long as it is, it is projecting. As long as it is, Dasein always has understood itself and always will understand itself in terms of possibilities' (1967: 185).

he died' (Lévi-Strauss 1963: 178). Our continuity and embeddedness in the world never come with guarantees.

On the coincidence of person and world, Willerslev offers a reading of Heidegger's human being as primordially absorbed in practical involvements in daily affairs that render self and world indistinguishable from one another. Willerslev's response to this is: 'If this were the case, then the experienced and the experiencer would conflate, would become one, thereby making any experience of the world impossible' (2007: 24). This falls back on an ontology of a knower-known relation which, as I have shown, Heidegger believes is *not* basic to human experience. In bringing to light the not-at-home, however, as part of an account of being-in-the-world, it should be clear that there is always something that can never be fully *there*, absorbed or accounted for in experience. Agamben, in Heideggerian terms, shows what is lacking in Willerslev's reading of *Being and Time*:

> What interests Heidegger here as a mark of factical experience is this dialectic of concealment and unconcealment, this double movement by which whoever wants to know everything while remaining concealed in knowledge is known by a knowledge that is concealed from him. Facticity is the condition of what remains concealed in its opening, of what is exposed by its very retreat. (Agamben 1999:106)

Not only is being never fully present or absorbed into the 'hurly-burly of human actions' (Wittgenstein 1967: 567); in the attempt to get clear of itself, human being is also always in part concealed in its own unconcealment, or unravelled in its own ravelling. It is here that the metaphysical perspective must produce a methodological and epistemic shift within anthropology which allows for antinomies, discontinuity, deadlocks and the absent to be just as insightful as the 'positivities' of presence, continuity and subsumption. They are each caught up in the other.

CONCLUSION

Phenomenological anthropologists have rightly stressed the necessity of persons' being-in-the-world. Now, however, I believe we should inquire into the *nature* of the necessity of the relation between world and persons. Do anomie, madness, disease, and failures of praxis occur as part of 'being-in-the-world' or as part of the radical experiential discontinuity of the 'not-at-home'? My thesis is that our practical, everyday involvements in the world comprise the metaphysical grounds of being a person, but not its effects. The labouring of everyday life to secure a form of existence for oneself, family, friends, or whoever they may be, is necessarily tied with and overlaps the lives of others. The countless situations when, as persons and anthropologists, we are faced with life-projects which are struggling, however, suggests a greater dynamism to being-in-the-world, since it is never an 'account' that is settled, so to speak. Being-in-the-world does not mean one cannot fall out of the world; we are never held so securely.

As I have suggested, by thinking of falling-out and being-in, being-at-home-in-the-world and not-being-at-home-in-the-world, entangling and disentangling, as part of the same life-movements and not as opposite poles, we can begin to give richer accounts of how persons make their way in the world. Disease and death do not need to be thought of solely as threats, nor does having a life with few obvious difficulties mean contentment and ease with the world. The world, though given to us in experience as the *given*, is nevertheless a *given* which can be taken, or simply fall, away.

When discussing the case of Adel with several friends I was struck by how, each in their own way, they all had a pre-thematized understanding of how every life was reliant on others and how insecure these reliances could be. One particular friend recounted the last weeks of his mother's life when it was all too clear that she had not long to live. She had fought for over a year but during those last days she had, as my friend said, 'loosened her grip on life'. Persons themselves can also disentangle themselves or simply let go and fall away when the struggle becomes too much. Through this strange intermingling of disparate intentionalities – our own and those of others – life emerges in forms we are all too rarely able to recognize in our everyday world.

As an indirect homage to the difficulties that Adel had to endure working in the hotel, several replacements have come and gone, unable to work the long hours for the low wage. Adel's labour is marked both physically and existentially by the inability of those others to maintain the struggling productivity that Adel had kept up for more than three years. Being-in-the-world is as much about our ability to maintain a grip on the social meshwork as it is about our slipping and disentangling. I would not want to end on too definitive a note and maybe the last word, as the first, belongs to Ingold: 'We are neither coming nor going but perpetually "coming and going"' (Ingold and Vergunst 2008:17).

GISLI PALSSON

12

Retrospect

> The phenomenon of life negates the boundaries that customarily divide our disciplines and fields.
>
> Hans Jonas (2001 [1966]: xxiv), *The Phenomenon of Life.*

Concerns with the relative unity and diversity of the discipline of anthropology have a long history. As early as 1936 Boas remarked that for a few decades physical and social anthropology had been drifting 'more and more apart' (Boas 1940: 172), a theme repeated three decades later in Alland's introduction to what he called 'Darwinian anthropology', separating 'the cultural traits which develop . . . against the background of biologically determined human nature' (Alland 2009: 236). Despite repeated recapturing of anthropology in recent decades (see, for example, Fox 1991, Segal and Yanagisako 2005), the two theoretical tectonic plates of biology and physiology, on the one hand, and, on the other, culture and society remain more or less intact. European biological and social anthropologists, for instance, continue to operate on very different terrains in the disciplinary landscape. In the North American context, there are recent cases of departmental fission along these lines, defying the classical four-field approach. As Schultz remarks, to challenge (and by the same token, to defend) the four-field approach is simply to 'muddy the waters' (2009: 227), avoiding the real issue of the reality of the two-field divide which splits the study of humans more generally, in practically every academic environment on the planet. Almost universally, anthropology is taught and practised as the dual field of the social and

Biosocial Becomings: Integrating Social and Biological Anthropology, eds T. Ingold and G. Palsson. Published by Cambridge University Press.

biological, variously constituted and articulated in relation to other fields, depending on historical and institutional context.

A series of recent intellectual developments have converged to challenge dominant thinking on the disciplinary structures of anthropology and, more generally, the divide between the life sciences and the humanities, fashioning a novel approach to the development and evolution of humans and other species. It seems time now to draw upon these developments to outline new avenues for the study of humans and explore opportunities for their advance, moving beyond mere critiques of dualisms and the anthropological 'mirage' of space between nature and nurture identified by Fox Keller (2010). Such an enterprise, the contributors to this book suggest, both entails a 'biosocial' rethinking of boundaries and a processual, relational perspective on becoming. Ingold has been a persistent key figure in critical anthropological engagement with 'biology', and more than anybody else he has informed and inspired this project in theoretical terms (Ingold, 1990, Chapter 1, this volume). In what follows, I will highlight what I see as some of the broad themes and implications of this collective enterprise.

INTELLECTUAL PASTURES AND DISCIPLINARY BOUNDARIES

Humans are endlessly engaged in some kind of boundary work, in all places at all times; indeed, the theme of the persistence of boundaries and their uses and abuses has a long and lively career in anthropology. The compartmentalization of the discipline itself and continued discussions of it, perhaps, need not surprise anthropologists. Some anthropologists, among them Shore (2010), have addressed the sites for the production of knowledge and the making of boundaries as ethnographic objects. Disciplines and subdisciplines are established and maintained – disciplined. 'Disciplining', as Clifford remarks (2005: 24), 'is not only a matter of defining scholarly territories, research topics, and analytic methods – the "content" of a discipline. The term evokes older traditions of normative training and ascetic practice that take modern form in pastoral and governmental institutions, including the university'. Spatial metaphors have typically been used with respect to such fragmentation; thus, the language of zoning, fields, pastures and territories. As Greenblatt emphasizes, the university is characterized by 'what might be called the rigid compartmentalization of mobility. Although in the past twenty years or so many academic disciplines have formally embraced ideas of "cultural mobility", they have for the most part operated with tunnel vision: the times and places in which they see significant mobility occurring remain strictly limited; in all other contexts, they remain focused on fixity' (2010: 2–3). Anthropology itself is a case in point. While the arrival of the new genetics has complicated the story in several ways on both sides of the great disciplinary divide (Goodman, Heath and Lindee 2003, Palsson 2007a, 2008, Marks 2011), forcing us to rethink nature and nurture in a novel fashion, the overall structures remain remarkably persistent.

It is necessary, however, to historicize the rigid language of intellectual pastures. As Foucault showed with his 'archaeology' of knowledge (1970), academic disciplines are recent and unstable phenomena established and modified in particular historical contexts. Disciplines, subdisciplines and fields of enquiry arise, develop and disappear, much like natural species. This evolutionary process begs important questions: What establishes the candidacy for a new field and how is it disciplined? Exploring the history of several fields, Lloyd emphasizes the 'distinctively inhibiting effects' of disciplinary boundaries (Lloyd 2009: 3). He discusses the role of elites claiming specialist knowledge in some particular field, professionalizing their subject, pointing out that elites are not always engines for progress:

> One of the recurrent phenomena we have encountered to a greater or lesser degree in all our individual studies is that of an elite jealously guarding its privileges – even, it may claim, its mandate – in the matter of how the subject should be pursued. The effect of elite influence may sometimes be not to stimulate and provide a framework for further development, but rather to restrict it, not to encourage innovation but to block it. (Lloyd 2009: 175)

One school of thought that has been gaining elite status as a hegemonic paradigm is the broad combination of evolutionary genetics and evolutionary psychology (see, for example, Mace, Holden and Shennan 2005, Kronfeldner 2006, Pagel 2012). In such approaches, the concepts of culture and society have been co-opted both for the purpose of fine-tuning neo-Darwinian theory and for extending its application to everything human. Culture, as the saying goes, evolves much like the biological world, through 'natural' selection and random variation. For some, the appeal to Darwinian theory is simply useful analogy, but for others it suggests ontological continuity. Alland argues that analogue models of evolution 'applied to culture fail because they ... deny continuity between biological and cultural processes, the very thing which one would expect from a more general theory of evolution' (2009: 267–268). Either way, while neo-Darwinism has generated interesting discussions and insights, appealing to and nourishing public sensibilities, in its elite capacity it has tended to resist more radical and innovative perspectives on the nature–culture divide and to block out alternative theorizing.

In a well-known passage in *The Origin of Species*, Darwin presented a notion of connectedness and interpenetration resolutely at odds with both his main thesis about the autonomy and role of species and current neo-Darwinian thought:

> It is interesting to contemplate an entangled bank clothed with many plants of many kinds, with birds singing in the bushes, with various insects flitting about, and with worms crawling through the damp earth, and to reflect that these elaborately constructed forms, so different from each other, and dependent on each other in so complex a manner, have all been produced by laws acting around us. ... There is grandeur in this view of life, with its several powers ... (Darwin 2008: 210–11)

Despite the grandeur of this view of entangled life and 'multiple agency', to use modern jargon (see, for instance, Bennett 2010), Darwin, along with other classical thinkers, would establish the powerful idea that living nature was essentially an assembly of distinct species of organisms, subject to the laws of change and evolution.

While neo-Darwinian theory has had an enormous impact and a broad disciplinary appeal, it increasingly seems a straitjacket. There is no need to invoke tired debates on social construction and the postmodern. Organisms and species increasingly seem indistinct and fuzzy – in Darwin's words 'entangled'; the environment is literally *embodied* in humans through social practices and epigenetic processes and, moreover, humans themselves partly *consist of* microbial ecosystems (Benezra, DeStephano and Gordon 2012). What is important now is to broaden the notion of Anthropos to allow for such indistinction and fuzziness, to recapture Darwin's vision of entanglement and to see where it would lead and what it would mean for anthropology and related disciplines, evolutionary theory, biopolitics and environmental practice.

A perspective on life as entangled, then, is not anti-Darwinian, although it seems fundamentally anti-*neo*-Darwinian. It is essential to go beyond anthropocentric terms such as 'competition' and 'cooperation' which have characterized neo-Darwinian theory in several fields. Margulis and Sagan propose 'a new search in the social sciences for terms to replace the old, tired social Darwinist metaphors' (2002: 15). The 'acquiring of genomes', they argue, has been central to the evolutionary process throughout. While the notion of the acquisition of genomes is a significant step forward, highlighting the flaws of methodological individualism in evolutionary theory, it is still focused on genes and genomes as both the driving forces of evolution and its 'frozen' embodied outcomes, disregarding epigenetic biosocial processes and other evidence that complicates the story of life. The entanglements of life are both vertical and horizontal and, moreover, social as well as biological, however one defines these terms. In light of this, it seems that the perspective of biosocial becomings is one of the most promising avenues on the theoretical agenda.

THE 'BIOLOGICAL' AND THE 'SOCIAL': MOVING BATTLE-LINES

One of the major dividing lines of the academy and its disciplinary architecture and classification, if not *the* dividing line, is that represented by the separation of the natural and the social. Nature has often been presented as one half of a pair – natural/social, biological/artificial, and so on. This is reflected in some earth-system notions that are fundamentally dualistic, 'linking' and 'coupling' the two systems of the earth and humans as if they were different realities (Berkes and Folke 1998). As Schabas points out in her analysis of the conceptual roots of economic thought, both natural and social scientists 'have implicitly agreed to divide the world into two parts. When physicists today think of the world they investigate, it is one with all the social institutions stripped away. . . . Economists, in parallel fashion, have come to adopt a domain of discourse that is similarly segregated. . . . To be sure, some of the analytical tools resemble one another . . . But there is no ontological conviction that economic phenomena are themselves part of the natural order that serves as the domain for natural scientists' (Schabas 2005: 12). More importantly, given the context of this book, the nature/culture divide remains firmly established *within* anthropology.

The separation of nature and society underlies the theoretical frameworks of evolutionary psychology and related neo-Darwinian developments, all of which rest on the fundamental notions of 'natural' selection and variation neatly separated from the social world. Ellen suggests that 'it is difficult to see how something called anthropological theory, which presumably seeks to answer the large and central questions about human distinctiveness and diversity, can be properly anthropological without some engagement with (rather than submission to) the biological' (2010: 292). To engage with the biological without submitting to it, however, is not necessarily to take it for granted (Lewontin 1983, Ingold 1990, 2011). Fraser, Kember and Lury pose questions 'about the role and status of (thinking) life across the natural sciences, the social sciences and humanities' in order to facilitate 'consideration of what might be included in the life sciences, how such sciences might be conducted, and who conducts them' (2005: 1).

For Ellen, Darwinian theory has more general relevance than any other framework: 'the only theory that has been sufficiently robust to resist all changes as an overarching framework', he argues, 'is some or other version of Darwinism relying on selection for reproductive fitness' (2010: 396). He opts for what he calls a 'twin-track mode', 'to treat culture simultaneously not only as non-genetic information transmitted between biological individuals and its material manifestations, but also as some complex Geertzian network of language-mediated symbols' (2010: 399). Levinson similarly maintains that human evolution 'has been shaped by the development of *two distinct* types of information transfer across generations, genetic and cultural, with systematic interactions between them' (2009: 26; my emphasis). Such a perspective, he suggests, 'makes us think differently about what the biological endowment for language must be. . . . On this account, the essential properties of language are divided between two inheritance systems, biological and cultural, and the long-term interactions between them' (Levinson 2009: 27–28). More broadly, Laland, Odling-Smee and Myles draw attention to 'gene–culture interactions', emphasizing that 'little consideration has been given to the possibility that cultural practices have transformed the selection pressures acting on humans' (2010: 137). What happens, one may wonder, to the evolutionary paradigm if the genetic and cultural realms not only interact but fundamentally overlap, if both selection and variation in the Darwinian scheme are thoroughly social?

The key point is not to insist upon a particular articulation of nature and society or to erase once and for all the ontological distinction itself – no doubt, some intellectual fields and communities will continue to revise it and find it useful for various purposes. The nature/society distinction, after all, is not only of ontological concern; it is also a serious practical and economic issue. Such a distinction, for instance, is often highlighted by legal discourse in the context of patent law. In 2010, a US federal judge invalidated patents related to two human genes, BRCA1 and BRCA2, whose mutations have been associated with breast cancer. The judge reasoned that the patents held by Myriad Genetics and the University of Utah Research Foundation were 'improperly granted' as they involved a 'law of nature', falling outside of the realm of things that can be patented (Schwartz and Pollack 2010). The relevant legal distinction is not between living and inanimate things, but between products of nature and human inventions. Thus, while patent laws usually do not embrace laws of nature or physical phenomena (the law of gravity and the Northern Lights, for example), *engineered* DNA molecules – as opposed to 'isolated DNA' which retains the same form as in the human body – are human-made inventions eligible for patent claims and, possibly, protection.

With the development and application of biotechnologies involving both cellular materials and 'artificial' machines, the space for 'pure' nature is progressively reduced. Both parties to the quarrels between natural science and humanities, Rheinberger notes, 'would do well to keep their sights trained on the moving battle-lines being drawn at the frontiers of research, for the boundary between nature and culture is constantly in flux there, if it is not, indeed, disappearing altogether' (2010: 4). This necessarily complicates the discourse on property and patent law. As Strathern observes, these developments invite concerns over 'aspects of life and body' which will loom large on the global agenda over the next decade and, as a result, one can expect 'an explosion of concern with ownership' (1998: 216).

Whatever the relative proportions and the boundaries of the natural and the artificial in legal contexts, however, what matters most of all to anthropological and biological theory is the move beyond essence and being to relations and becomings. As Schultz observes, 'relational evolutionary perspectives could ... help shift discussion away from sterile nature–nurture debates' (2009: 234). The significance of relationality, as Fraser, Kember and Lury argue more broadly in their critique of social scientific and biological reductionism, 'is that it acts as a "lure for life", an enticement to move beyond the conflation of life with the (life) sciences, to conceive life as not confined to living organisms, but as movement, a radical becoming' (2005: 3). 'The existence of changing distinctions, of ontological differentiations between nature and culture', they continue, is 'an effect of particular relations. ... Thinking in terms of process ... recognizes the *relations* between the traditions that privilege culture, on the one hand, or nature, on the other.'

TEMPORALITY: THE DURABILITY OF DISPOSITIONS

Dualism has been heavily criticized in anthropology in recent years (Descola and Palsson 1996, Ingold 2000, Hastrup 2013), partly drawing upon earlier critiques (Bateson 1972). This is not to say, however, that it has disappeared from the scene. Not only is it generally taken for granted among those who identify themselves with the 'biological' side of the academy; it is also sometimes sternly defended by both social anthropologists and the human sciences in general. Newton suggests that while dualism has often been presented 'as the evil "sin" to which any right minded social thinker should object' (2010: 200), the sheer frequency of anti-dualist sentiment within western thought indicates that there is no and has never been any hegemonic dualism to challenge or deconstruct. In his view, 'anti-dualist critique tends to downplay possible differences between "human society" and the rest of nature' (2010: 201). The key difference, he argues, between the biological, physical and social aspects of nature is that of timescales: 'Even the more pressing cases for social constancy appear "fleeting" when viewed from the millennial timescapes which we associate with some physical and biological processes' (Newton 2010: 204). 'In sum', Newton suggests, 'the *historical longevity* of certain biological processes and outcomes presents a marked contrast to the *historical specificity* of the social domain. Biological and social temporality is therefore not simply equivalent even though we exist within a bio-social terrain that remains nothing if not intimately intertwined' (2010: 205, original emphasis). The argument about the *longue durée* of the natural world and the relentless flow of structures and events in the social world is not convincing, however. After all, we should not underestimate the *constancy* of cultures, traditions, ethnic groups and languages and the persistence of the boundaries that they erect and maintain.

Moreover, during the current age of the so-called Anthropocene, practically everything 'natural' on planet Earth is undergoing change as a result of human impact. It now seems clear that 'natural' climate has much to do with 'artificial' human services and goods – and it is rapidly changing. Some volcanic eruptions might even be attributed to human activities; a recent study of Iceland's Vatnajökull ice cap suggests that melting glaciers can increase volcanic activity over timeframes that are relevant to humans (Pagli and Sigmundsson 2008). This is geopolitics in a new sense and on a new scale (Palsson, Szerszynski, Sörlin *et al.* 2013). Not only do human activities affect the natural world at the giga-scale of the planet, the biological constitution of organisms is increasingly informed by both deliberate human impact (genetic engineering, for instance) and the unintended side effects of human activities. Thus, growing awareness of environmental epigenetic processes independent of the genetic code necessarily complicates the classic differentiation between 'nature's handiwork' and the 'artificial' products of human creativity.

Society and environment not only enter the body, they also have a lasting impact. As Niewöhner argues, it is ironic that the more intensively biologists probe into the human body the less they are able to ignore the environment within which it is embedded: 'Environmental epigenetics produces an "embedded body", that is, a body that is heavily impregnated by its own past and by the social and material environment within which it dwells. It is a body that is imprinted by evolutionary and transgenerational time, by "early-life", and a body that is highly susceptible to changes in its social and material environment' (2011: 290). Non-DNA-related aspects of our developmental trajectory, then, turn out to be inherited from one generation to another. These constitutional transformations or dispositions, to use the language of Bourdieu, may be just as durable as those attributed to genomic sequences. This is biosocial becoming in a very real sense, the Anthropocene at the nano-scale. The modern village green is not just the global environment but our bodies as well. Nature, whether at the level of cells, organisms, ecosystems, or the planet, turns out to be just as 'fleeting' as society, undermining any attempt to separate the two analytical and theoretical domains in terms of different timescales.

Inevitably, the recent escalation of human impacts on the rest of the natural world has sensitized many people to biosocial temporality at the giga-scale, at the planetary level, undermining, for instance, the classic distinction between the relative permanence of climate and the transitory nature of weather. At the same time, theories of both materiality and sociality in the social sciences and archaeology have emphasized the durability and embodiment of the structures of everyday life. Social relations, Sofaer suggests, are 'biologically embedded' (2006: 77). In the practice theory of Bourdieu, social life is above all the result of human agency, a continuous process of becoming necessarily embodied in the form of the *habitus*, 'the durably installed generative principle of regulated improvisations' (1977: 78). For Bourdieu, the notion of embodied 'dispositions' plays a central role: 'It expresses first the *result of an organizing action*, with a meaning close to that of words such as structure; it also designates a *way of being, a habitual state* (especially of the body) and, in particular, a *predisposition, tendency, propensity*, or *inclination*' (1977: 214, n. 1; original emphasis). Humans, then, make history, but their making is both impelled and retarded by the biosocial habitus, the durable dispositions embodied through past historical experience.

ALTERNATIVE THEORETICAL LANGUAGES?

It seems time now to explore alternative theoretical languages in relation to life, what they would entail for epistemic frameworks and academic practice, and how they might allow us to take a step forward. The two key terms in this direction, 'naturecultures' and 'biosociality' – popularized, respectively, by Haraway (2008), Rabinow (1996) and some others – have slightly different trajectories, etymologies and connotations. To simplify a highly nuanced landscape of theoretical language, while the former draws upon cultural studies and theories of culture highlighting semiotics, meaning and 'superorganic' symbolic frameworks – suspensions 'in webs of significance', in Geertz's terms (1973: 5) – the latter draws upon social theory, emphasizing sociality and the social constitution of being. Whatever their differences, the terms 'biosociality' and 'naturecultures' seem equally suitable for addressing both relational aspects and processes of becoming. Indeed, they tend to be applied to similar phenomena and to respond to more or less the same theoretical problems focusing on the fuzzy boundaries of biology and society in the contemporary world.

The biosocial has a long history, gaining popularity already in the 1970s. Steiner's *After Babel* referred to the 'bio-social' aspects of language to highlight its individuality: 'The fact that tens of thousands of different, mutually incomprehensible languages have been or are being spoken on our small planet is a graphic expression of the deeper-lying enigma of human individuality, of the bio-genetic and bio-social evidence that no two human beings are totally identical' (Steiner 1975: 49–50). Similarly, Paredes *et al.* spoke of the 'adaptive biosocial mechanisms' of the Tarahumara Indians of Mexico, to 'biosocial parameters such as body build, physical capacity, aggressive behavior and alcohol preference' (1970: 163). In the Tarahumara context, 'Community feeling, uniformity of culture and effective communication are achieved through the institution of *tesguinada*, a periodic celebration in which groups of men and women work and drink together forming overlapping social networks which constitute the molecules of their social body' (Paredes *et al.* 1970: 173). Such usages of the biosocial still seem interactive, assuming two separate systems that nevertheless somehow inform each other.

Recent applications of the biosocial and biosociality go further, highlighting the need to overcome the nature/culture split in the light of the refashioning of biological material through the new genetics, along with the experimentation and discourse entailed in the 'rewriting' of 'life itself'. In the course of these applications, the meaning of biosociality has somewhat narrowed, to refer to social identities and groupings based on genetic information about biological ancestry and analyses of medical risks and conditions (Hacking 2006). For Gibbon and Novas, the concept of biosociality is useful for exploring the 'biologies and socialities that are assembled by a range of practitioners and social actors across a variety of interconnected sites' (2007: 1). While primarily focused on identity issues, this kind of conceptualization points in the direction of both a relational perspective and a framework of becoming.

HUMAN VARIATION AND BOUNDARIES: SELF AND OTHER

One of the issues that tend to separate social and biological anthropology is that of 'human variation'. In anthropological parlance, this usually refers to variation in anatomy or biology, not to the social and cultural variance documented, for instance, by comparative ethnography. While human variation is one of the defining themes of physical and biological anthropology, it is not as pure, a-social or precultural as one might think. It is often assumed, by both scientists and the public at large, that biology is a property somehow unevenly distributed among groups, as if some populations were more biological than others. Lindee and Ventura Santos point out that the 'frequently uninterrogated notion that some people are "more" biological, whatever that might mean, has played a role in scientific constructions of race, gender, class and ethnicity for at least two centuries' (2012: 7). Often the people who work with such notions also assume that they themselves, and others of 'their kind', are more cultural.

For decades, such scaling of biological and cultural difference was established and illustrated through bone measurements, in particular by means of the 'anthropometron' and the 'craniometer' of the last couple of centuries. Racial signatures, classifications and typologies were said to be engraved in the biological world, not imposed by the spectator reading the measurements. Physical anthropology carried an aura of proper, authentic science. Another early claim to scientific exploration of human variation was that of fingerprint analysis. Sir Francis Galton suggested that fingerprints were 'in some respects the most important of all anthropological data. We shall see that they form patterns, of a curious variety of shape, whose boundaries can be firmly outlined, and which are little worlds in themselves' (Galton 1892: 1–2), one of the advantages being that each set of fingerprints 'differentiates the person who made it, throughout the whole of his life, from all the rest of mankind' (1892: 4). Galton also hoped to show that fingerprints varied with populations, with what he called 'racial differences', a project that he concluded was a failure. While fingerprint analyses continue to be extensively used in forensics, here too they have a mixed reputation. Some studies portray fingerprint forensics as a subjective enterprise characterized by 'cultish demeanor' (Ashbaugh, quoted in Epstein 2002: 645).

With the development of molecular biology, DNA gradually replaced skeletal material as *the* avenue into human history and variation. This would quickly complicate the issues of identity and variation. One of the most stunning discoveries of genomic studies concerns the ease with which genes may be routinely transferred between organisms of different kinds. 'Alien' microbes, it turns out, constitute most of the cells and the genes in the human body. Recent studies of the human microbiome draw a picture of a body stacked with trillions of bacteria, one or more kilos in adults (The Human Microbiome Project Consortium 2012). The microbes play a variety of important roles, keeping us healthy – facilitating the digestion of food and the synthesis of vitamins – and making us sick.

Including the microbiome as a fundamental part of the body, and not just as an appendix to a 'host', suggests that we need to revise and expand our notions of humans. As a result, the biological approach to human variation is likely to change. Benezra, DeStephano and Gordon speak of an 'anthropology of microbes', emphasizing that 'investigating microbes from an ethnographic perspective should provide anthropologists with new perspectives about how human biology and social practices are inextricable', allowing us to develop a concept of the 'biological-social self' (2012: 1). At the same time, the inclusion of the microbiome also poses a challenge to evolutionary theory. As Margulis and Sagan argue in their discussion of 'miscegenous' genomes, the 'acquisition of the reproducing other, of the microbe and the genome, is no mere sideshow. Attraction, merger, fusion, incorporation, co-habitation, recombination ... and other forms of forbidden couplings, are the main sources of Darwin's missing variation' (2002: 205). If humans are assemblies or aggregates of life forms, the outcomes of ensembles of biosocial relations (Palsson, Chapter 2, this volume), then they have not simply co-evolved *with* more-than-human microbes; humans *are* microbes, in a literal sense. In light of this, some scholars have suggested that humans should be seen as 'holobionts' (Dominguez-Bello and Blaser 2011: 453), discrete units of living matter ('bionts') involving both human and microbial cells.

A further challenge to the notions of the autonomous human and of vertical ancestry, of self and other, is that of gestational chimerism, the intermingling during pregnancy of cells from two or more genetically distinct bodies. In the orthodox view of human reproduction, the human fetus and the mother are two distinct individuals with distinct evolutionary objectives, maintaining the status of 'all self' during the process of gestation. Chimerism was classically described by Walknowska in 1969, in a study that identified fetal cells with Y chromosomes circulating in the blood of women pregnant with male fetuses, but it was only decades later that the broad implications, including philosophical and evolutionary ones (Howes 2008), began to be explored (see Kelly 2012). It seems that a woman who has been pregnant is likely to be a chimera, acquiring 'foreign' cells and genomes from every pregnancy. Her body, in other words, is potentially inscribed with each pregnancy, whether or not it resulted in birth, and hence the presence of some genetic material from the father in her body. This highlights troubled body boundaries or interfaces between sexual partners, siblings and across generations.

Drawing upon Haraway's work (1991) on biology as the offspring of political discourse and Martin's work (1990) on the immune system as a key site in western biopolitics focused on the establishment of self, other and boundaries, often through metaphors of a defended nation-state, Kelly argues that the integrity of these troubled boundaries 'is intimately tied to contemporary western understandings of the individual subject and associated forms of autonomy, relationality and community' (Kelly 2012: 13). Chimeric evidence then, raises questions about genetic intermingling and entanglement, diverging from current understandings, biological and social, of selves as autonomous worlds, as purely self. Chimerism is not only important for the understanding of the 'natural' maternal-fetal interface, it is also increasingly relevant for the framing of life in the wake of therapeutic developments based on experimental manipulation, the transfer of cells and molecules through permeable bodies by means of tissue transplants and regenerative medicine. The maternal–fetal interface, Kelly argues (2012: 11), 'may be the site of transmission of cells from one sibling to another (horizontal transmission) as well as between generations (vertical transmission). According to these interpretations of evidence, the boundaries between bodies are messy indeed!' The neo-Darwinian theory of modern evolutionary biology, along with sociobiology and evolutionary psychology, which introduced notions such as 'inclusive fitness' and 'kin selection' in order to explain cooperative behaviour and altruism, now have to vastly expand their horizons to include chimeric cell transfer and the laterally transferred genes of trillions of bacteria – a daunting task.

If natural selection and chance variation are necessarily thoroughly biosocial, neo-Darwinian theory seems to lose much of its robustness. It seems pertinent to evoke Bakhtinian notions of language: 'language, for the individual consciousness, lies on the borderline between oneself and the other. The word in language is half someone else's. It becomes "one's own" only when the speaker populates it with his own intention, his own accent, when he appropriates the word, adapting it to his own semantic and expressive intention' (Bakhtin 1981: 293). Life is the noisy communication of biosocial assemblies with a plurality of speakers and voices, not a neat back-and-forth over-to-you dialogue modelled on early telephone conversations. The neo-Darwinian legacy, after all, turns out to be an uncomfortable intellectual straitjacket, incapable of addressing and incorporating such complexities. A truly vigorous take on human life and human variation must integrate the social and the biological, the individual and the collective, ontogeny and phylogeny, organism and context, being and becoming.

LIFE AS LIVED AND BECOMING

Recently, in the wake of the new genetics, several social theorists (see, for instance, Rabinow 1996, Franklin and Lock 2003, Rose 2006) have focused on what has been rendered as 'life itself' or 'bare life'. Drawing upon the 'biopolitical' works of Foucault, these scholars have usefully drawn critical attention to the life sciences, in particular the radical refashioning of health and living matter and its broad impact on social life. As Fassin has pointed out, however, Foucault's biopolitics 'is not a politics of life as the etymology would suggest but a politics of population understood as a community of living beings: "life" remains largely elusive' (Fassin 2009: 46). Canguilhem, one of Foucault's influential teachers, distinguished between two notions of life – the living (*le vivant*) and the lived (*le vécu*) – but recent works on the politics of life itself, including those of Foucault, seem to focus only on the former, the living minus the lived, 'life which is lived through a body (not only through cells) and as a society (not only as species)' (Fassin 2009: 48). For 'the lived' Fassin proposes the label 'life as such', to avoid a narrow focus on 'biological' phenomena, the 'forms' of life, and the reduction of living beings to populations at the expense of lived life.

Foucault's notion of 'pastoral power' as a paradigm of biopolitics, Fassin points out, draws upon the metaphor of the shepherd, 'in charge not of individuals but of a herd – or a flock of souls' (2009: 54). Such a notion is helpful for addressing the politics and power games of states, regimes and empires informed by modern biomedicine, the use of population statistics – what Hacking (1982: 281) calls 'the avalanche of numbers' – and the progressive extension of the scientific gaze into bodies, cells and genomes. In order, however, to capture the phenomenology of life (life as such, as lived), to refashion the life sciences, and to integrate the two anthropologies of the social and the biological, another perspective is needed.

Canguilhem's *Knowledge of Life*, not surprisingly, offers some hints. Unlike Foucault, he remained fairly committed to the lived, emphasizing worlds of becoming and the pragmatic experience of the organism, including humans:

> The use of concepts and intellectual tools forged by . . . the biologist, in order to understand the experience of life proper to the organism is . . . at once both inevitable and artificial. We will . . . say that the knowledge of life must take place through unpredictable conversions, as it strives to grasp a becoming whose meaning is never so clearly revealed to our understanding as when it disconcerts it. (Canguilhem 2008: 22)

Having been expelled from nature through the Darwinian framing of evolution as sheer, unpredictable adventure, the pragmatics of purpose and becoming need to be reintroduced into accounts of the unfolding of life. 'Whatever their success so far', Jonas wrote in 1966, 'all contemporary revisions of the traditional ontology indeed start, almost axiomatically, from the conception of being as *becoming*, and in the phenomenon of cosmic evolution look for the key to a possible stand beyond the old alternatives' (Jonas 2001: 58; original emphasis).

A theory of biosocial becomings needs to draw upon such insights and many others recently advanced in the same vein, highlighting a move from states of being and discrete things with given essences and identities to movement, relations and processes. Humans *become* human through relations with other becoming organisms and species and the environments within which they are embedded.

Informed by agency and communication, these relations are often seen as restricted to humans – and possibly a small set of honorary non-humans such as other 'higher' primates, cetaceans, dogs and certain birds. Anthropologists have often imagined that they could safely cling to the cliché that only humans are capable of tool-making, inference and subjectivity. It is difficult, however, to see why biosocial becoming should only pertain to a select few. Each of the traditional indicators that were once presumed to separate human culture from the natural world of non-human animals has proved to be flawed or far more complex or ambivalent in its application than previously thought. Animals in general, it is often pointed out nowadays, are capable of acting, relating, sensing and suffering. Connolly wonders if such rethinking implies erasing all differences between the natural and cultural: 'Do natural processes contain as much agency as cultural processes? An amoeba as much as a human? . . . No. But it does encourage us to rethink the dicey problematic of agency, to convert a dichotomous view that bestows agency upon humans only – or in many cases upon humans to some degree and God to an infinite degree – into a more *distributive* image of agency' (2011: 22). Some anthropologists have called for an expanded, 'multispecies' ethnography (Kirksey and Helmreich 2010), moving beyond the traditional order of Primates with which many biologically minded anthropologists operate, to document and theorize interspecies relations.

While the community or the *polis* of the lived is often extended to 'non-human' animals, partly through the activities of the animal rights movement, the rest of organic life is invariably excluded as if it simply and docilely belonged to 'the environment', devoid of agency, relationality and communication. Drawing upon Deleuze and Guattari, Houle (2011) has argued that we should pay serious attention to plant community, signalling and communication. Her point is not just to expand the community even further, to *extend* the notions of rights and ethics beyond animals to herbal life, but to allow for some form of becoming, the possibility of what this volume identifies as biosocial becoming. Houle's work aims to show that vegetal modalities call for genuinely different ways of thinking, 'rather than nifty *vegetal-variation on* our dominant modes of enacting communication and our dominant ways of thinking about what communication is and is in the service of' (2011: 98; original emphasis). Studies of plant behaviour and the chemical profiles of their signals, Houle suggests, force us 'to imagine not only that plants are individuals, but that these individuals are continuously co-evolving *with*, and *in*, varying environmental relations which themselves are evolving in complex ways' (2011: 100).

The notion of 'ethology' – usually limited to animals – may thus have a much broader basis and relevance than normally imagined. The results of research on plant sensibility and behaviour are often mind-boggling (Chamovitz 2012). One of the perennial problems is that the scientific study of plant signalling seems necessarily to 'isolate and fix its samples (genetically, geographically, temporally), and to carry-on "as if in a common garden".... What is lacking is *the living matrix itself*' (Houle 2011: 110: my emphasis). Houle seems to introduce the Bakhtinian notion of speech mentioned above to the life of plants, emphasizing noisy communication, fluid communities and a plurality of voices: 'The punchline is that the teloi or "self-realization" of plant communication is neither strictly individual nor even species specific but is accomplished in and through radical kinships, through a fantastically versatile and multidirectional capacity to harmonize a multiplicity of actions. Whatever plants are up to, it *is* complex being-together in the world, an *original sociality* going beyond any simple sense of *between*' (Houle 2011: 111; original emphasis). In light of this, the application of the duality of nature and culture to the living matrix begins to look exceedingly restrictive. Plants, too, are 'wired for culture', in the language of evolutionary psychology (Pagel 2012). Seamless biosocial becoming seems to apply universally to the phenomenon of life.

TOWARDS AN ANTHROPOLOGY OF LIFE

Over the last hundred years or so, anthropology has taken radical turns in terms of institutional foundations, theoretical framing and professional practice inside as well as outside the academy. Yet the disciplinary tensions along the nature/society divide are as strong as ever. On the 'natural' side, in the wake of advances in genetics and genomics, biological anthropology has largely taken the lead from physical anthropology, sometimes aligned with evolutionary psychology, evolutionary biology and related schools. To the extent that culture and society have been addressed in these fields, they have been seen as separate realms, the results of natural selection, sometimes 'shaping' the human genome by transforming selective pressures (Laland, Odling-Smee and Myles 2010). On the other side of the tectonic divide, sociocultural anthropology has similarly reified and bounded its own subject matter and theoretical terrain, with minimal attention to the biological and the material. Strangely enough for the study of a single species, life itself and life as such seem to belong to separate intellectual galaxies. Important realigning and theoretical reframing, however, should not be ignored.

The ethnography of the twenty-first century seems destined to reframe the relations of humans and the environment in non-reductionist terms, emphasizing the porous boundaries and relational nature of human bodies (see, for instance, Niewöhner 2011). This is in stark contrast to the 'ecological anthropology' developed in the 1960s under the influence of the theory of 'cultural ecology' and related themes. The pioneers of the ethnography of ecological anthropology, to be sure, left for their field sites armed with new questions about human life and its material context, drawing attention to environmental issues that had long been more or less ignored by social and cultural anthropology. Rappaport's *Pigs for the Ancestors* (1968), for instance, carefully explored relations among humans and pig populations, their representations in local 'cognized' models, and their ritual regulation. However, while this scholarship was a leap forward, paving the way for a new generation of anthropologists and a series of innovative ethnographies focusing on environmental relations and concerns, in hindsight it had significant limits. For one thing, the human body remained a black box, unexplored and under-theorized, reaffirming the disciplinary divide between biology and society. For another, the relations of humans and other animals were narrowly reduced to matters of human subsistence and energy transfer.

The evidence on chimera, microbiomes and epigenetics now calls for an extended, entangled notion of the Anthropos, other animals and life in general. As Barnes and Dupré note, 'it is striking how easily molecular and genomic perspectives suggest ways of ordering the biological world profoundly opposed to those built into the most widely diffused and accepted accounts of evolution. Since the time of the "New Synthesis" these accounts have been very strongly atomistic, both in their Darwinist view of the individual organism and their Mendelian view of the unit of inheritance, whether gene or allele' (2008: 139). Methodological individualism has no place in modern theorizing on life, human or otherwise.

Given the relational notion of embedded bodies and the importance of interspecies relations and entanglements, the framework of biosocial becomings advanced here seems highly pertinent, for integrating social and biological anthropology and for relating both to the general pursuit of the phenomenon of life. In her historical account of anthropology, Smocovitis points out that this is 'the only ... disciplinary category allying itself as part of the life sciences devoted to the study of one species' (2012: 122). While we do have disciplines focused on a class of species – 'primatology' and 'ornithology', for instance – we do not speak of 'drosophilology', nor would we have two separate fields focusing on, respectively, the biology and culture of fruit flies: 'Clearly, the existence of anthropology in general and biological anthropology in particular', Smocovitis concludes,

> in some measure still preserves the special or 'unique' status given to humans despite the fact that humans are animals. From a proper biological perspective, this is of course profoundly anthropocentric, as we might expect of an area devoted exclusively to humans. Given this etymological confusion, should we perhaps rethink the meaning of the category of 'anthropology', so heterogeneous that it has become the locus for interdisciplinarity, as something even more inclusive but amorphous, such as 'human studies'? (Smocovitis, 2012: 122)

It is important to explore what such rethinking of the discipline would imply. Perhaps the study of humans is necessarily anthropocentric in some respects. After all, we are studying ourselves and, as a result, we lose an element of objectivity and a clear sense of species differences, similarities and relations. Despite the grand disciplinary rhetoric concerning the importance of a comparative perspective in the study of humans, this is, by definition, anthropology *completely* at home in a deep sense (Palsson 2007a: 27); members of a particular species accumulating knowledge about their own species-being, through their own involvement with conspecifics, using the senses and sensibilities of their own minds and bodies. Anthropology, however, should be capable of relaxing some of its anthropocentric assumptions. Kohn suggests that the study of humans is inseparable from the study of other species and, as a result, it might be redefined as the study of more than one species, as the 'anthropology of life' (2007: 6). Recently, there have been several calls along these lines, to address relations and processes of entanglement, and to consider 'the interconnected *becoming* of life in its more-than-human form' (Panelli 2010: 82).

The framework of biosocial becomings draws upon strands in classical philosophy and social theory. Marx argued, for example, that humans are ensembles or assemblies fundamentally embedded in an 'inorganic body', namely nature, and that history is, in consequence, part of nature. 'History itself', he insisted, 'is a real part of natural history, and of nature's becoming man' (Marx 1975: 355). At the same time, the biosocial-becomings perspective launched here links together a variety of recent theoretical developments in several fields, all of which facilitate rethinking of the relations and processes of life and its becoming. Such a broad perspective should not be seen as a fixed baseline or an end in itself but as a starting point for further work, as a tentative framework inviting novel conceptual and theoretical development and elaboration.

While the authors of the chapters in this volume converge on certain issues, they no doubt disagree on others, drawing upon different theoretical traditions and applying their conceptual tools in different ethnographic and analytical terrains. The point is not to silence other voices or to establish a new elite, but to open up an innovative space for theoretical developments that take the merging of the biological and the social for granted, moving beyond conventional conceptual and disciplinary boundaries that seem stagnated and restrictive. While we may abandon the tidy notion of two systems and the fragmenting of anthropology, we should also heed Connolly's warning: 'Participation in a world of becoming teaches modesty about the powers of argument even while appreciating its pertinence' (2011: 10). Nevertheless the integration of social and biological anthropology, and the realignment of the social and the biological in theoretical terms, seem more timely than ever, drawing as they do upon a family resemblance of several recent strands of thought which seek to provide a better understanding of the Anthropos and of life as lived, as such, in the contemporary world. The anomalies and contradictions of orthodox models of biological and cultural evolution are not inconsequential. They must be tackled head-on.

References

Abuelas de Plaza de Mayo. (2008). *Las Abuelas y la Genética: El Aporte de la Ciencia en la Búsqueda de los Chicos Desaparecidos*. Buenos Aires: Abuelas de Plaza de Mayo.

Abuelas de Plaza de Mayo. (2012). www.abuelas.com.ar. (Last accessed 18 June).

Aessopos, A., M. Kati and J. Meletis. (2007). Thalassemia intermedia today: should patients regularly receive transfusions? *Transfusion*, **47**: 792–800.

Agamben, G. (1999). The passion of facticity. In *Potentialities: Collected Essays in Philosophy*. Stanford, CA: Stanford University Press, pp. 185–204.

Agamben, G. (2004). *The Open: Man and Animal*, trans. K. Attell. Stanford, CA: Stanford University Press.

Al-Mohammad, H. (2010). Towards an ethics of being-with: intertwinements of life in post-invasion Basra. *Ethnos*, **75**: 425–446.

Alberts, J. R. (2008a). The nature of nurturant niches in ontogeny. *Philosophical Psychology*, **21**: 295–303.

Alberts, J. R. (2008b). Self-sensitivity in fetal development. *Infancy*, **13**: 270–274.

Alland, A. (2009 [1967]). *Human Evolution and Human Behavior: An Introduction to Darwinian Anthropology*. London: Routledge.

Antonow-Schlorke, I., M. Schwab, L. A. Cos *et al.* (2011). Vulnerability of the fetal primate brain to moderate reduction in maternal global nutrient availability. *Proceedings of the National Academy of Sciences USA*, **108**: 3011–3016.

Appadurai, A. (1996) *Modernity at Large: Cultural Dimensions of Globalization*. Minneapolis, MN: University of Minnesota Press.

Archer, J. (2004). Sex differences in aggression in real world settings: a meta-analytic review. *Review of General Psychology*, **8**: 291–322.

Ardener, E. (1989). Social anthropology and population. In *The Voice of Prophecy and Other Essays*, ed. M. Chapman. Oxford: Blackwell, pp. 109–127.

Arditti, R. (1999). *Searching for Life: The Grandmothers of the Plaza de Mayo and the Disappeared Children of Argentina*. Berkeley, CA: University of California Press.

Arendt, H. (1958). *The Human Condition*. Chicago, IL: University of Chicago Press.

Argento, A. (2009). *De vuelta a casa {Historias de hijos y nietos restituidos}*. Buenos Aires: Marea.

Ariew, A. (2006). Innateness. In *Handbook of the Philosophy of Science*, eds. M. Matthen and C. Stevens. Amsterdam: Elsevier, pp. 1–18.

Aristotle. (2009). *The Nicomachean Ethics*. Oxford: Oxford University Press.

Armelagos, G. and A. Goodman. (1998). Race, racism and anthropology. In *Building a New Biocultural Synthesis: Politic-Economic Perspectives on Human Biology*, eds. A. Goodman and T. Leatherman. Ann Arbor, MI: University of Michigan Press, pp. 359–378.

Avital, E. and E. Jablonka. (2000). *Animal Traditions: Behavioural Inheritance in Evolution*. Cambridge: Cambridge University Press.

Aymemi, A. (1942). *Los Bubis en Fernando Poo*. Madrid: Imprenta de Galo Saez.

Bahuchet, S. (1992). *Dans la forêt d'Afrique Centrale: Les pygmées aka et baka*. Louvain: Peeters.

Bainbridge, D. (2003). *Making Babies: The Science of Pregnancy*. Cambridge, MA: Harvard University Press.

Bakhtin, M. M. (1981). *The Dialogic Imagination: Four Essays*, ed. M. Holquist, trans. C. Emerson and M. Holquist. Austin, TX: University of Texas Press.

Bamford, S. (2004). Conceiving relatedness: non-substantial relations among the Kamea of Papua New Guinea. *Journal of the Royal Anthropological Institute*, **19**: 287–306.

Barker, D. J. P. (1995). Fetal origins of coronary heart disease. *British Medical Journal*, **31**: 171–174.

Barker, D. J. P. (2004). The developmental origins of adult disease. *Journal of the American College of Nutrition*, **23**: 588S–595S.

Barkow, J. H., L. Cosmides and J. Tooby (eds.) (1992). *The Adapted Mind: Evolutionary Psychology and the Generation of Culture*. New York, NY: Oxford University Press.

Barnard, A. (1992). *Hunters and Herders of Southern Africa: A Comparative Ethnography of the Khoisan Peoples*. Cambridge: Cambridge University Press.

Barnard, R. J., N. Kobayashi and W. J. Aronson. (2008). Effect of diet and exercise intervention on the growth of prostate epithelial cells. *Prostate Cancer and Prostatic Diseases*, **11**: 362–366.

Barnes, B. and J. Dupré. (2008). *Genomes and What to Make of Them*. Chicago. IL: University of Chicago Press.

Barrett, L., R. Dunbar and J. Lycett. (2002). *Human Evolutionary Psychology*. Princeton, NJ: Princeton University Press.

Barton, D. and M. Hamilton. (2005). Literacy, reification and the dynamics of social interaction. In *Beyond Communities of Practice: Language, Power, and Social Context*, eds. D. Barton and K. Tusting. Cambridge: Cambridge University Press, pp. 14–35.

Barua, B. P. (2001). *Ethnicity and National Integration in Bangladesh: A Study of the Chittagong Hill*. New Delhi: Har Anand Publications.

Basso, K. H. (1976). "Wise words of the western Apache": metaphor and semantic theory. In *Meaning in Anthropology*, eds. K. H. Basso and H. A. Selby. Albuquerque, NM: University of New Mexico Press, pp. 93–123.

Bateson, G. (1972). *Steps to an Ecology of Mind*. Frogmore: Paladin.

Bateson, G. (1979). *Mind and Nature: A Necessary Unity*. Cresskill, NJ: Hampton Press.

Bateson, P. (1988). The active role of behavior in evolution. In *Evolutionary Processes and Metaphors*, eds. M. W. Ho and S. W. Fox. London: John Wiley and Sons, pp. 191–207.

Bateson, P. (2004). The origins of human differences. *Daedalus*, **133**: 36–46.

Bateson, P. (2005). The return of the whole organism. *Journal of Biosciences*, **30**: 31–39.

Benezra, A., J. DeStephano and J. I. Gordon. (2012). Anthropology of microbes. *Proceedings of the National Academy of Sciences USA* **109**: 6378–6381.

Bennett, J. (2010). *Vibrant Matter: A Political Ecology of Things*. Durham, NC: Duke University Press.

Benson, S. (2006). Injurious names: naming, disavowal, and recuperation in contexts of slavery and emancipation. In *The Anthropology of Names and Naming*, eds. G. vom Bruck and B. Bodenhorn. Cambridge: Cambridge University Press, pp. 177–199.

Berkes, F. and C. Folke (eds.) (1998). *Linking Social and Ecological Systems: Management Practices and Social Mechanisms for Building Resilience*. Cambridge: Cambridge University Press.

Beynon, H. (1973). *Working for Ford*. London: Penguin Books.

Bitkeeva, A. N., V. Y. Gusev, O. A. Povoroznyuk *et al.* (2005). *Endangered Languages of Indigenous Peoples of Siberia* [Online]. Available at: http://lingsib.iea.ras.ru/en/ [Accessed 15 September 2011].

Blackmore, S. (2000). *The Meme Machine*. Oxford: Oxford University Press.

Bloch, M. (1998). *How We Think They Think: Anthropological Studies in Cognition, Memory and Literacy*. Boulder, CO: Westview Press.

Bloch, M. (2005). *Essays on Cultural Transmission*. Oxford: Berg.

Bloch, M. (2007). Durkheimian anthropology and religion: going in and out of each other's bodies. In *Religion, Anthropology, and Cognitive Science*, eds. H. Whitehouse and J. Laidlaw. Durham, NC: Carolina Academic Press, pp. 63–83.

Bloch, M. (2011). The blob. *Anthropology of this Century*, **1** (May 2011). Available at: http://aotcpress.com [Accessed October 2011].

Boas, F. (1916). Tsimshian mythology. In *Thirty-first Annual Report of the Bureau of Ethnology*. Washington, DC: Smithsonian Institution.

Boas, F. (1940). *Race, Language and Culture*. New York, NY: Macmillan Company.

Bodenhorn, B. (2000). "He used to be my relative": exploring the bases of relatedness among the Inupiat of Northern Alaska. In *Cultures of Relatedness: New Approaches to the Study of Kinship*, ed. J. Carsten. Cambridge: Cambridge University Press, pp. 128–148.

Bodenhorn, B. and G. vom Bruck. (2006). "Entangled in histories": an introduction to the anthropology of naming. In *The Anthropology of Names and Naming*, eds. G. vom Bruck and B. Bodenhorn. Cambridge: Cambridge University Press, pp. 1–30.

Bogin, B. (1999 [1988]). *Patterns of Human Growth*. New York, NY: Cambridge University Press.

Bourdieu, P. (1972). *Esquisse d'une théorie de la pratique précédé de Trois études d'ethnologie kabyle*. Paris: Éditions du Seuil.

Bourdieu, P. (1977). *Outline of a Theory of Practice*. Cambridge: Cambridge University Press.

Bourdieu, P. (2000). *Pascalian Meditations*. Stanford, CA: Stanford University Press.

Bozzola, M., P. Travaglino, N. Marziliano *et al.* (2009). The shortness of Pygmies is associated with severe under-expression of the growth hormone receptor. *Molecular Genetics and Metabolism*, **98**: 310–313.

Bradley, R. H. and R. F. Corwyn. (2002). Socioeconomic status and child development. *Annual Review of Psychology*, **53**: 371–399.

Brentano, F. C. (1995[1874]). *Psychology from an Empirical Standpoint*. London: Routledge.

Bruchac, J. (1995). *An Anthology of Contemporary Cherokee Prose*. New York, NY: Greenfield Review Press.

Burawoy, M. (1979). *Manufacturing Consent: Changes in the Labor Process under Monopoly Capitalism*. Chicago, IL: University of Chicago Press.

Buss, D. M. (1994). *The Evolution of Desire*. New York, NY: Basic Books.

Bygren, L. O., G. Kaati and S. Edvinsson. (2001). Longevity determined by paternal ancestors' nutrition during their slow growth. *Acta Biotheoretica*, **49**: 53–59.

Cambrosio, A., A. Young and M. Lock. (2000). Introduction. In *Living and Working With the New Medical Technologies: Intersections of Inquiry*, eds. M. Lock, A. Young and A. Cambrosio. Cambridge: Cambridge University Press, pp. 1–16.

Canguilhem, G. (2008 [1965]). *Knowledge of Life*. New York, NY: Fordham University Press.

Cao, A. and R. Galanello. (2010). Beta-thalassemia. *Genetics in Medicine*, **12**: 61–76.

Cappellini, M. D., K. M. Musallam and A. T. Taher. (2009). Insight onto the pathophysiology and clinical complications of thalassaemia intermedia. *Hemoglobin*, **33** (S1): S145–S159.

Cardwell, C. R., L. C. Stene, G. Joner *et al.* (2008). Caesarean section is associated with an increased risk of childhood onset type 1 diabetes mellitus: a meta-analysis of observational. *Diabetologica*, **51**: 726–735.

Castoriadis, C. (1988). *Political and Social Writings, volume 2, 1955–1960: From the Workers' Struggle Against Bureaucracy to Revolution in the Age of Modern Capitalism*, trans. D. A. Curtis. Minneapolis, MN: University of Minnesota Press.

Cavell, S. (1976). *Must We Mean What We Say? A Book of Essays*. Cambridge: Cambridge University Press.

Chaiklin, S. and J. Lave. (1996). *Understanding Practice: Perspectives on Activity and Context*. Cambridge: Cambridge University Press.

Chamovitz, D. (2012). *What a Plant Knows: A Field Guide to the Senses*. New York, NY: Scientific American/Farrar, Strauss and Giroux.

Chapais, B. (2008) *Primeval Kinship: How Pair-Bonding Gave Birth to Human Society*. Cambridge, MA: Harvard University Press.

Charney, E. (2012). Behavior genetics and post genomics. *Behavioral and Brain Sciences*, **35**: 331–358.

Charpentier, M. J. E., R. C. Van Horn, J. Altmann and S. C. Alberts. (2008). Paternal effects on offspring fitness in a multimale primate society. *Proceedings of the National Academy of Sciences USA*, **105**: 1988–1992.

Chatjouli, A. (2012). *Thalassemiked zoes. Biologiki diafora, kanonikotita, biokinonikotita. Mia anthropologiki proseggisi* [Thalassaemic Lives. Biological Difference, Normality and Biosociality. An Anthropological Approach]. Athens: Patakis.

Cherkas, L. F., A. Aviv, A. M. Valdés *et al.* (2006). The effects of social status on biological aging as measured by white-blood-cell telomere length. *Aging Cell*, **5**: 361–365.

Clark, H. H. and S. E. Brennan. (1991). Grounding communication. In *Perspectives on Socially Shared Cognition*, eds. L. B. Resnick, J. M. Levine and S. D. Teasley. Washington, DC: American Psychological Association, pp. 127–149.

Clarke, A. E., J. K. Shim, L. Mamo, J. R. Fosket and J. R. Fishman. (2003). Biomedicalization: technoscientific transformations of health, illness, and U.S. biomedicine. *American Sociological Review*, **68**: 161–194.

Clifford, J. (2005). Rearticulating anthropology. In *Unwrapping the Sacred Bundle: Reflections on the Disciplining of Anthropology*, eds. D. A. Segal and S. J. Yanagisako. Durham, NC: Duke University Press, pp. 24–48.

Conklin, B. (2001). *Consuming Grief: Compassionate Cannibalism in an Amazonian Society*. Austin, TX: Texas University Press.

Connolly, W. (2011). *A World of Becoming*. Durham, NC: Duke University Press.

Cooper, R., C. Rotimi, S. Ataman *et al.* (1997). The prevalence of hypertension in seven populations of West African origin. *American Journal of Public Health*, **87**: 160–168.

Cormier, L. A. (2003). *Kinship with Monkeys: The Guajá Foragers of Eastern Amazonia*. New York, NY: Columbia University Press.

Cosmides, L. and J. Tooby. (1994). Beyond intuition and instinct blindness: toward an evolutionary rigorous cognitive science. *Cognition*, **50**: 41–77.

Critchley, S. (2002). Enigma variations: an interpretation of Heidegger's *Sein und Zeit*. *Ratio*, **15**: 154–175.

Csordas, T. (1990). Embodiment as a paradigm for anthropology. *Ethos*, **18**: 5–47.

Csordas, T. (1993). Somatic modes of attention. *Cultural Anthropology*, **8**: 135–156.
Danziger, E. (2006). The thought that counts: interactional consequences of variation in cultural theories of meaning. In *Roots of Human Sociality: Culture, Cognition and Interaction*, eds. N. Enfield and S. Levinson. Oxford: Berg, pp. 259–278.
Darwin, C. (2008 [1859]). *Evolutionary Writings*, ed. J. A. Secord. Oxford: Oxford University Press.
Darwin, C. (1872). *On The Origin of Species* (6th edition). London: John Murray.
Darwin, F. (ed.) (1887). *The Life and Letters of Charles Darwin, Including an Autobiographical Chapter*. London: John Murray.
Davenport, S., P. Johnson and Yuwali. (2005). *Cleared Out: First Contact in the Western Desert*. Canberra: Aboriginal Studies Press.
Dawkins, R. (1976). *The Selfish Gene*. Oxford: Oxford University Press.
Dawkins, R. (1982). *The Extended Phenotype*. Oxford: Oxford University Press.
Dawkins, R. (1986). *The Blind Watchmaker*. Harlow, Essex: Longman Scientific & Technical.
Day, R. L., K. N. Laland and F. J. Odling-Smee. (2003). Rethinking adaptation: the niche-construction perspective. *Perspectives in Biology and Medicine*, **46**: 80–95.
Deleuze, G. and F. Guattari. (2004). *A Thousand Plateaus: Capitalism and Schizophrenia*, trans. B. Massumi. London: Continuum.
DeMallie, R. J. (2009). Community in native America: continuity and change among the Sioux. *Journal de la société des américanistes*, **95**: 185–206.
Descola, P. (1996). Constructing natures. In *Nature and Society. Anthropological Perspectives*, eds. P. Descola and G. Palsson. London: Routledge, pp. 82–102.
Descola, P. (2005). *Par-delà nature e culture*. Paris: Gallimard.
Descola, P. and G. Palsson (eds.) (1996). *Nature and Society: Anthropological Perspectives*. London: Routledge.
Dittmer, K. (1961). *Die sakralen Häuptlinge der Gurunsi im Obervolta-Gebiet Westafrica*. Hamburg: Cram, De Gruyter und Co.
Dobzhansky, T. (1951 [1937]). *Genetics and the Origins of Species*. New York, NY: Columbia University Press.
Dominguez-Bello, M. G. and M. J. Blaser. (2011). The human microbiota as a marker for migrations of individuals and populations. *Annual Review of Anthropology*, **40**: 451–474.
Dreyfus, H. L. (1991). *Being-In-the-World: A Commentary on Heidegger's Being and Time, Division I*. Cambridge, MA: MIT Press.
du Boulay, J. (1984). The blood: symbolic relationships between descent, marriage, incest prohibitions, and spiritual kinship in Greece. *Man* **19**: 533–556.
Dunbar, R., C. Gamble and J. Gowlett. (2010). *Social Brain, Distributed Mind. Proceedings of the British Academy 158*. New York: Oxford University Press.
Dunbar, R., R. Baron, A. Frangou *et al.* (2012). Social laughter is correlated with an elevated pain threshold. *Proceedings of the Royal Society B, Biological Sciences*, **279** (1731): 1161–1167.
Duranti, A. (2006). The social ontology of intentions. *Discourse Studies*, **8**: 31–40.
Duranti, A. (2008). Further reflections on reading other minds. *Anthropological Quarterly*, **81**: 483–494.
Durham, W. H. (1991). *Coevolution: Genes, Culture and Human Diversity*. Stanford, CA: Stanford University Press.
Eckert, P. (2007). *Pitjantjatjara Yankunytjatjara Picture Dictionary*. Alice Springs: IAD Press.
Einsiedel, E., A. Allansdottir, A. Chatjouli *et al.* (2002). Brave new sheep – the clone named Dolly. In *Biotechnology: The Making of a Global Controversy*, eds.

M. W. Bauer and G. Gaskell. Cambridge: Cambridge University Press, pp. 313–347.

El-Hani, C. N. (2007). Between the cross and the sword: the crisis of the gene concept. *Genetic Molecular Biology*, **30**: 297–307.

Elango, N. and S. V. Yi. (2008). DNA methylation and structural and functional bimodality of vertebrate promoters. *Molecular Biology and Evolution*, **25**: 1602–1608.

Elder, G. H. (1974). *Children of the Great Depression*. Chicago, IL: Chicago University Press.

Ellen, R. (2010). Theories in anthropology and "anthropological theory". *Journal of the Royal Anthropological Institute*, **16**: 387–404.

Elliot, C. and P. Brodwin. (2002). Identity and genetic ancestry tracing. *British Medical Journal*, **325**: 1469–1471.

Ellison, P. T. and P. B. Gray (eds.) (2009). *The Endocrinology of Social Relationships*. Cambridge, MA: Harvard University Press.

ENCODE Project and Consortium. (2004). The ENCODE (Encyclopedia of DNA Elements) Project. *Science*, **306** (5696): 636–640.

ENCODE Project and Consortium. (2007). Identification and analysis of functional elements in 1% of the human genome by the ENCODE pilot project. *Nature*, **447** (7146): 799–816.

ENCODE Project and Consortium. (2012). An integrated encyclopedia of DNA elements in the human genome. *Nature*, **489**: 5–170.

Endicott, K. (1979). *Batek Negrito Religion: The World-View and Rituals of a Hunting and Gathering People of Peninsular Malaysia*. Oxford: Clarendon Press.

Enfield, N. (2006). Social consequences of common ground. In *Roots of Human Sociality: Culture, Cognition and Interaction*, eds. N. Enfield and S. Levinson. Oxford: Berg, pp. 399–430.

Enfield, N. and S. Levinson (eds.) (2006). *Roots of Human Sociality: Culture, Cognition and Interaction*. Oxford: Berg.

Epstein, R. (2002). Fingerprints meet *Daubert*: the myth of fingerprint 'science' is revealed. *Southern California Law Review*, **75**: 605–657.

Erikson, Ph. (1996). *La griffe des aïeux: marquage du corps et démarquages ethniques chez les Matis d'Amazonie*. Louvain/Paris: Peeters.

Erikson, Ph. (2000). The social significance of pet-keeping in Amazonia. In *Companion Animals and Us*, eds. E. Podberscek, E. Paul and J. Serpell. Cambridge: Cambridge University Press, pp. 7–26.

Erikson, Ph. (2004) *Qu'est-ce qu'un 'ethnonyme'?* [Online]. Available at: http://alhim.revues.org/index112.html [Accessed 27 September 2011].

Esteller, M., P. G. Corn, S. B. Baylin and J. G. Herman. (2001). A gene hypermethylation profile of human cancer 1. *Cancer Research*, **61**: 3225–3229.

Estruch, R., M. A. Martínez-González, D. Corella *et al.* (2006). Effects of a Mediterranean-style diet on cardiovascular risk factors: a randomized trial. *Annals of Internal Medicine*, **145**: 1–11.

Fabian, J. (1983). *Time and the Other: How Anthropology Makes Its Object*. New York, NY: Columbia University Press.

Fairhead, J. and M. Leach. (1996). *Misreading the African Landscape: Society and Ecology in a Forest–Savanna Mosaic*. Cambridge: Cambridge University Press.

Fassin, D. (2009). Another politics of life is possible. *Theory, Culture and Society*, **26**: 44–60.

Fausto, C. (1999). Of enemies and pets: warfare and shamanism in Amazonia. *American Ethnologist*, **26**: 933–956.

Fausto-Sterling, A. (2005). The bare bones of sex: Part 1, sex and gender. *Signs: Journal of Women in Culture and Society*, **30**: 1491–1528.

Fausto-Sterling, A. (2008). The bare bones of race. *Social Studies of Science*, **38**: 657–694.

Fernandez, J. (1974). The mission of metaphor in expressive culture. *Current Anthropology*, **15**: 119–145.

Fernandez-Reyes, D., A. G. Craig, S. A. Kyes *et al.* (1997). A high frequency African coding polymorphism in the N-terminal domain of ICAM-1 predisposing to cerebral malaria in Kenya. *Human Molecular Genetics*, **6**: 1357–1360.

Fienup-Riordan, A. (1991). *The Real People and the Children of Thunder: The Yup'ik Eskimo Encounter with Moravian Missionaries John and Edith Kilbuck*. Norman, OK: University of Oklahoma Press.

Fienup-Riordan, A. (2000). *Hunting Traditions in a Changing World: Yup'ik Lives in Alaska Today*. Rutgers, NJ: Rutgers University Press.

Finch, J. (2008). Naming names: kinship, individuality and personal names. *Sociology*, **42**: 709–725.

Flack, J. C., M. Girvan, F. B. M. de Waal and D. C. Krakauer. (2006). Policing stabilizes construction of social niches in primates. *Nature*, **439**: 426–429.

Flinn, M. and B. England. (1997). Social economics of childhood glucocorticoid stress response and health. *American Journal of Physical Anthropology*, **102**: 33–53.

Ford, E. S., W. H. Giles and W. H. Dietz. (2002). Prevalence of the metabolic syndrome among US adults: findings from the third National Health and Nutrition Examination Survey. *Journal of the American Medical Association* **16**: 356–359.

Foucault, M. (1970). *The Order of Things*. London: Routledge.

Fox, R. G. (1991). *Recapturing Anthropology: Working in the Present*. Santa Fe, NM: School of American Research Press.

Fox Keller, E. (2005). The century beyond the gene. *Journal of Bioscience*, **30**: 3–10.

Fox Keller, E. (2010). *The Mirage of a Space Between Nature and Nurture*. Durham, NC: Duke University Press.

Fraga, M. F., E. Ballestar, M. F. Paz *et al.* (2005). Epigenetic differences arise during the lifetime of monozygotic twins. *Proceedings of the National Academy of Sciences USA*, **102**: 10604–10609.

Franklin, S. (2003). Re-thinking nature-culture: anthropology and the new genetics. *Anthropological Theory*, **3**: 65–85.

Franklin, S. (2006). The cyborg embryo: our path to transbiology. *Theory, Culture and Society*, **23**: 167–187.

Franklin, S. and M. Lock (eds.) (2003). *Remaking Life and Death: Towards an Anthropology of the Biosciences*. Santa Fe, NM: School of American Research Press.

Fraser, M., S. Kember and C. Lury. (2005). Inventive life: approaches to the New Vitalism. *Theory, Culture and Society*, **22**: 1–14.

The Free Dictionary. (2010). http://www.thefreedictionary.com/. [Accessed 24 June 2010].

Fry, D. (2007). *Beyond War: The Human Potential for Peace*. New York, NY: Oxford University Press.

Fuentes, A. (2009a). *Evolution of Human Behavior*. New York, NY: Oxford University Press.

Fuentes, A. (2009b). A new synthesis: resituating approaches to the evolution of human behavior. *Anthropology Today*, **25**: 12–17.

Fuentes, A. (2010). The new biological anthropology: bringing Washburn's new physical anthropology into 2010 and beyond. *Yearbook of Physical Anthropology*, **53**: 2–12.

Fuentes, A. (2011). *Biological Anthropology: Concepts and Connections* (2nd edition). New York, NY: McGraw-Hill Publishing Company.

Fuentes, A. (2012). *Race, Monogamy, and Other Lies They Told You: Busting Myths of Human Nature*. Berkeley, CA: University of California Press.

Fuentes, A. (2013). Cooperation, conflict, and niche construction in the genus *Homo*. In *War, Peace, and Human Nature*, ed. D. Fry. New York, NY: Oxford University Press, pp. 78–94.

Fuentes, A., M. Wyczalkowski and K. C. MacKinnon. (2010). Niche construction through cooperation: a nonlinear dynamics contribution to modeling facets of the evolutionary history in the genus *Homo*. *Current Anthropology*, **51**: 435–444.

Fuller, D. (2005). *Writing the Everyday: Women's Textual Communities in Atlantic Canada*. Montreal: McGill–Queen's University Press.

Fullilove, M. T. (2009). Toxic sequelae of childhood sexual abuse. *American Journal of Psychiatry* **166**: 1090–1092.

G. 291. XLIII Gualtieri Rugnone de Prieto, Emma Elidia y otros s/sustracción de menores de 10 años. Argentine Supreme Court [online]. Available at: www.jcs.org.ar [Accessed 18 June 2012].

Galanello, R. and R. Origa. (2010). Beta-thalassemia. *Journal of Rare Diseases*, **5**: 11.

Galton, F. (1892). *Finger Prints*. London: Macmillan and Co.

Gare, A. E. (1995). *Postmodernism and the Environmental Crisis*. London: Routledge.

Geertz, C. (1973). *The Interpretation of Culture*. New York, NY: Basic Books.

Georges, E. (2008). *Bodies of Knowledge: The Medicalization of Reproduction in Greece*. Nashville, IN: Vanderbilt University Press.

Gerstein, M. B., C. Bruce and J. S. Rozowsky. (2007). What is a gene, post-ENCODE? History and updated definition. *Genome Research*, **17**: 669–681.

Gibbon, S. and C. Novas. (2007). Introduction: biosocialities, genetics and the social sciences. In *Biosocialities, Genetics and the Social Sciences: Making Biologies and Identities*, eds. S. Gibbon and C. Novas. London: Routledge, pp. 1–18.

Gibson, J. J. (1979). *The Ecological Approach to Visual Perception*. Boston, MA: Houghton Mifflin.

Gingeras, T. R. (2007). Origin of phenotypes: genes and transcripts. *Genome Research*, **17**: 682–690.

Gingrich, A. (2010). Ethnologie. In *Evolution: Ein Interdisziplinäres Handbuch*, eds. P. Sarasin, T. P. Weber and M. Sommer. Stuttgart: Metzler, pp. 226–234.

Glass, A. and D. Hackett. (2003). *Ngaanyatjarra & Ngaatjatjarra to English Dictionary*. Alice Springs: IAD Press.

Gluckman, P. H., M. A. Hanson and K. L. Thornburg. (2008). Effect of in utero and early-life conditions on adult health and disease. *New England Journal of Medicine*, **359**: 61–73.

Goffman, E. (1959). *The Presentation of Self in Everyday Life*. New York, NY: Anchor Books.

Gold, R. (1995). Owning our bodies: an examination of property law and biotechnology. *San Diego Law Review*, **32**: 1167–1247.

Goldberg, A. D., C. D. Allis and E. Bernstein. (2007). Epigenetics: a landscape takes shape. *Cell*, **128**: 635–638.

Golovnev, A. V. and G. Osherenko. (1999). *Siberian Survival: The Nenets and Their Story*. Ithaca, NY: Cornell University Press.

Goodman, A. H. (2000). Why genes don't count (for racial differences in health). *American Journal of Public Health*, **90**: 1699–1702.

Goodman, A. H., D. Heath and M. S. Lindee (eds.) (2003). *Genetic Nature/Culture: Anthropology and Science beyond the Two-Culture Divide*. Berkeley, CA: University of California Press.

Goodman, A. H., R. B. Thomas, A. C. Swedlund and G. J. Armelagos. (2005). Biocultural perspectives on stress in prehistoric, historical, and contemporary population research. *American Journal of Physical Anthropology*, **31**: 169–202.

Goodwin, C. and A. Duranti. (1992). Rethinking context: an introduction. In *Rethinking Context: Language as an Interactive Phenomenon*, eds. A. Duranti and C. Goodwin. Cambridge: Cambridge University Press, pp. 1–42.

Goody, E. (ed.) (1995). *Social Intelligence and Interaction. Expressions and Implications of the Social Bias in Human Intelligence*. Cambridge: Cambridge University Press.

Gottlieb, G. (1971). Ontogenesis of sensory function in birds and mammals. In *The Biospsychology of Development*, eds E. L. Tobach, L. Aronson and E. Shaw. New York, NY: Academic Press, pp. 67–128.

Gottlieb, G. (1997). *Synthesizing Nature–Nurture: Prenatal Roots of Instinctive Behavior*. Hillsdale, NJ: Lawrence Erlbaum.

Gottlieb, G. (2001). A developmental psychobiological systems view: early formulation and current status. In *Cycles of Contingency: Developmental Systems and Evolution*, eds. S. Oyama, P. E. Griffiths and R. D. Gray. Cambridge, MA: MIT Press, pp. 41–54.

Gould, S. J. (1977). *Ontogeny and Phylogeny*. Cambridge, MA: Harvard University Press.

Gould, S. J. (2002). *The Structure of Evolutionary Theory*. Cambridge, MA: Belknap Press of Harvard University Press.

Gow, P. (2001). *An Amazonian Myth and Its History*. Oxford: Oxford University Press.

Greenblatt, S. (2010). Cultural mobility: an introduction. In *Cultural Mobility: A Manifesto*, ed. S. Greenblatt *et al.*. Cambridge: Cambridge University Press, pp. 1–23.

Griffiths, P. E. (2007). Evo-Devo meets the mind: toward a developmental evolutionary psychology. In *Integrating Evolution and Development*, eds. R. Brandon and R. Sansom. Cambridge, MA: MIT Press, pp. 195–225.

Gruzdeva, E. (1998). *Nivkh (Languages of the World)*. Newcastle: LINCOM Europa.

Guignon, C. B. (1983). *Heidegger and the Problem of Knowledge*. Indianapolis, IN: Hackett Publishing.

Gurevich, A. (1992). *Historical Anthropology of the Middle Ages*. Cambridge: Polity Press.

Hacking, I. (1982). Biopower and the avalanche of printed numbers. *Humanities in Society*, **5**: 279–295.

Hacking, I. (2006). Genetics, biosocial groups and the future of identity. *Daedalus*, **135**: 81–95.

Hall, B. K. (2000). Balfour, Garstang and de Beer: the first century of evolutionary embryology. *American Zoologist*, **40**: 718–728.

Hallowell, A. I. and J. S. H. Brown. (1992). *The Ojibwa of Berens River, Manitoba: Ethnography into History*. Fort Worth, TX: Harcourt Brace Jovanovich.

Hamilton, W. D. (1964). The genetical evolution of social behaviour. I and II. *Journal of Theoretical Biology*, **7**: 1–16 & 17–52.

Hammoudi, A. (1997). *Master and Disciple: The Cultural Foundations of Moroccan Authoritarianism*. Chicago, IL: University of Chicago Press.

Haraway, D. (1985). A manifesto for cyborgs: science, technology and socialist feminism in the 1980s. *Socialist Review*, **80**: 65–107.

Haraway, D. (1991). *Simians, Cyborgs and Women: The Reinvention of Nature*. New York, NY: Routledge.

Haraway, D. (2003). *The Companion Species Manifesto: Dogs, People, and Significant Otherness*. Chicago, IL: Prickly Paradigm Press.

Haraway, D. (2008). *When Species Meet*. Minneapolis, MN: University of Minnesota Press.

Hart, D. and R. W. Sussman. (2008). *Man the Hunted: Primates, Predators, and Human Evolution*. Boulder, CO: Westview.

Harvey, G. (2000). *Indigenous Religions: A Companion*. Bath: Bookcraft Ltd.

Hastrup, K. (ed.) (2013). *Anthropology and Nature*. London: Routledge.

Hayami, Y., A. Tanabe and Y. Tokita-Tanabe (eds.) (2003). *Gender and Modernity: Perspectives from Asia and the Pacific*. Kyoto: Kyoto University Press.

Heidegger, M. (1967 [1927]). *Being and Time*. Oxford: Blackwell.

Heidegger, M. (1985 [1936]). *Schelling's Treatise on the Essence of Human Freedom*. Ohio, OH: Ohio University Press.

Heidegger, M. (1998 [1949]). Introduction to 'What is metaphysics?'. In *Pathmarks*, ed. W. McNeill. Cambridge: Cambridge University Press, pp. 277–290.

Helmreich, S. (2009). *Alien Ocean: Anthropological Voyages in Microbial Seas*. Berkeley, CA: University of California Press.

Henare, A., M. Holbraad and S. Wastell (eds.) (2007). *Thinking Through Things*. London: Routledge.

Herzfeld, M. (1992). *The Social Production of Indifference: Exploring the Symbolic Roots of Western Bureaucracy*. Chicago, IL: University of Chicago Press.

Herzfeld, M. (1997). *Cultural Intimacy: Social Poetics in the Nation-State*. New York, NY: Routledge.

Heyward, N. (1963). *Sarawak, Brunei, & North Borneo*. Singapore: Eastern Universities Press.

Hill, A. V. S., C. E. M. Allsopp and D. Kwiatkowsi. (1991). Common West African HLA antigens are associated with protection from severe malaria. *Nature*, **352**: 595–600.

Hilliard, W. M. (1968). *The People in Between: The Pitjantjatjara People of Ernabella*. London: Hodder & Stoughton.

Ho, D. H. and W. W. Burggren. (2010). Epigenetics and transgenerational transfer: a physiological perspective. *Journal of Experimental Biology*, **213**: 3–16.

Ho, M.-W. and P. T. Saunders. (1979). Beyond Darwinism: an epigenetic approach to evolution. *Journal of Theoretical Biology*, **78**: 573–791.

Hornborg, A. (2001). *The Power of the Machine: Global Inequalities of Economy, Technology, and Environment*. Walnut Creek, CA: AltaMira Press.

Horner, J. (2004). *Seeking Racial Justice: An Insider's Memoir of the Movement for Aboriginal Advancement*. Canberra: Aboriginal Studies Press.

Houle, K. L. F. (2011). Animal, vegetable, mineral: ethics as extensions or becoming? The case of becoming-plant. *Journal of Critical Animal Studies*, **9**: 89–116.

Howell, S. (1984). *Society and Cosmos: Chewong of Peninsular Malaysia*. Oxford: Oxford University Press.

Howes, M. (2008). Self and nonself. In *A Companion to the Philosophy of Biology*, eds. S. Sarkar and A. Plutyanski. Oxford: Blackwell, pp. 271–286.

The Human Microbiome Project Consortium. (2012). A framework for human microbiome research. *Nature*, **486**: 215–221.

Hutchins, E. (1995). *Cognition in the Wild*. Cambridge, MA: MIT Press.

Hutchins, E. (2006). The distributed cognition perspective on human interaction. In *Roots of Human Sociality: Culture, Cognition and Interaction*, eds. N. Enfield and S. Levinson. Oxford: Berg, pp. 375–398.

Hyyppä, M. T., J. Mäki, O. Impivaara and A. Aromaa. (2006). Leisure participation predicts survival: a population-based study in Finland. *Health Promotion International*, **21**: 5–12.

Iacoboni, M. (2008). *Mirroring People: The New Science of how We Connect with Others*. New York, NY: Picador.

Ingold, T. (1986). *The Appropriation of Nature: Essays on Human Ecology and Social Relations*. Manchester: Manchester University Press.
Ingold, T. (1990). An anthropologist looks at biology. *Man* (N.S.), **25**: 208–229.
Ingold, T. (1991). Becoming persons: consciousness and sociality in human evolution. *Cultural Dynamics*, **4**: 355–378.
Ingold, T. (1996). Situating action V: the history and evolution of bodily skills. *Ecological Psychology*, **8**: 171–182.
Ingold, T. (2000). *The Perception of the Environment: Essays on Livelihood, Dwelling and Skill*. London: Routledge.
Ingold, T. (2001a). From complementarity to obviation: on dissolving the boundaries between social and biological anthropology, archaeology, and psychology. In *Cycles of Contingency: Developmental Systems and Evolution*, eds. S. Oyama, P. E. Griffith and R. D. Gray. Cambridge, MA: MIT Press, pp. 255–279.
Ingold, T. (2001b). From the transmission of representations to the education of attention. In *The Debated Mind: Evolutionary Psychology Versus Ethnography*, ed. H. Whitehouse. Oxford: Berg, pp. 113–153.
Ingold, T. (2004). Beyond biology and culture: the meaning of evolution in a relational world. *Social Anthropology*, **12**: 209–221.
Ingold, T. (2006a). Becoming persons: consciousness and sociality in human evolution. In *Anthropology in Theory: Issues in Epistemology*, eds. H. Moore and T. Sanders. Malden, MA: Blackwell.
Ingold, T. (2006b). Rethinking the animate, re-animating thought. *Ethnos*, **71**: 9–20.
Ingold, T. (2007a). *Lines: A Brief History*. Abingdon: Routledge.
Ingold, T. (2007b). Materials against materiality. *Archaeological Dialogues*, **14**: 1–16.
Ingold, T. (2007c). The trouble with "evolutionary biology". *Anthropology Today*, **23**: 13–17.
Ingold, T. (2008). Bringing things to life: creative entanglements in a world of materials. Paper presented at the conference *Vital Signs: Researching Real Life*, 9–11 September. Available at: http://www.reallifemethods.ac.uk/events/vitalsigns/programme/ingold.htm [Accessed 28 May 2011].
Ingold, T. (2009). Point, line and counterpoint: from environment to fluid space. In *Neurobiology of "Umwelt": How Living Beings Perceive the World*, eds. A. Berthoz and Y. Christen. Berlin: Springer-Verlag, pp. 141–155.
Ingold, T. (2010a). *Culture evolves? Not as you think*. Paper presented at workshop on *Beyond the Biological and the Social: Anthropology as the Study of Human Becomings*, at the 11th Biennial Conference of the European Association of Social Anthropologists, Maynooth, Ireland.
Ingold, T. (2010b). The textility of making. *Cambridge Journal of Economics*, **34**: 91–102.
Ingold, T. (2010c). The man in the machine and the self-builder. *Interdisciplinary Science Reviews*, **35**: 353–364.
Ingold, T. (2011). *Being Alive: Essays on Movement, Knowledge and Description*. Abingdon: Routledge.
Ingold, T. (2013). *Making: Anthropology, Archaeology, Art and Architecture*. Abingdon: Routledge.
Ingold, T. and J. Lee Vergunst. (2008). Introduction. In *Ways of Walking: Ethnography and Practice on Foot*, eds. T. Ingold and J. Lee Vergunst. Aldershot: Ashgate, pp. 1–20.
International Human Genome Sequencing Consortium. (2001). Initial sequencing and analysis of the human genome. *Nature*, **409**: 860–921.

Irving, A. (2004). Life made strange: an essay on the reinhabitation of bodies and landscapes. In *The Qualities of Time: Anthropological Approaches*, eds. W. James and D. Mills. Oxford: Berg, pp. 317–331.

Jablonka, E. and M. Lamb. (2002). The changing concept of epigenetics. *Annals of the New York Academy of Sciences USA*, **981**: 82–96.

Jablonka, E. and M. Lamb. (2005). *Evolution in Four Dimensions: Genetic, Epigenetic, Behavioral, and Symbolic Variation in the History of Life*. Cambridge, MA: MIT Press.

Jablonka, E. and G. Raz. (2009). Transgenerational epigenetic inheritance: prevalence, mechanisms, and implications for the study of heredity and evolution. *Quarterly Review of Biology*, **84**: 131–176.

Jackson, M. (1983). Knowledge of the body. *Man* (N.S.), **18**: 327–345.

Jackson, M. (1995). *At Home in the World*. Durham, NC: Duke University Press.

Jackson, M. (2004). *Existential Anthropology: Events, Exigencies, and Effects*. New York, NY: Berghahn.

Jackson, M. (2009). *The Palm at the End of the Mind: Relatedness, Religiosity, and the Real*. Durham, NC: Duke University Press.

James, W. (1979). *Kwanim Pa: The Making of the Uduk People*. Oxford: Clarendon Press.

Jasanoff, S. (ed.) (2004). *States of Knowledge: The Co-Production of Science and the Social Order*. London: Routledge.

Jeeves, M. (ed.) (2011). *Rethinking Human Nature: A Multidisciplinary Approach*. Cambridge: Eerdmans.

Joh, E. (2011). DNA theft: recognizing the crime of nonconsensual genetic collection and testing. *Boston University Law Review*, **91**: 665–700.

Johannsen, W. (1909). *Elemente der exakten Erblichkeitslehre*. Jena: Gustav Fischer.

Johannsen, W. (1911 [1910]). The genotype conception of heredity. *American Naturalist*, **45**: 129–159.

Jonas, H. (2001 [1966]). *The Phenomenon of Life*. Evanston, IL: Northwestern University Press.

Just, R. (1989). Triumph of the ethnos. In *History and Ethnicity*, eds. E. Tonkin, M. Chapman and M. McDonald. London: Routledge, pp. 71–88.

Kaati, G., L. O. Bygren and S. Edvinsson. (2002). Cardiovascular and diabetes mortality determined by nutrition during parents' and grandparents' slow growth period. *European Journal of Human Genetics*, **10**: 682–688.

Kavanagh, K. D., A. R. Evans and J. Jernvall. (2007). Predicting evolutionary patterns of mammalian teeth from development. *Nature*, **449**: 427–432.

Keller, L. (1999). *Levels of Selection in Evolution*. Princeton, NJ: Princeton University Press.

Kelly, S. E. (2012). The maternal-foetal interface and gestational chimerism: the emerging importance of chimeric bodies. *Science as Culture*, **1**: 1–25.

Kendal, J., J. J. Tehrani and J. Odling-Smee. (2011). Human niche construction in interdisciplinary focus. *Philosophical Transactions of the Royal Society B: Biological Sciences*, **366**: 785–792.

Kim-Cohen, J., T. E. Moffitt, A. Caspi and A. Taylor. (2004). Genetic and environmental processes in young children's resilience and vulnerability to socioeconomic deprivation. *Child Development*, **75**: 651–668.

King, R. (1978). In the sanding booth at Ford. In *Social Problems in Today's World*, eds. J. Perry and E. Perry. Boston: Little, Brown, pp. 199–205.

Kirksey, E. and S. Helmreich. (2010). The emergence of multispecies ethnography. *Cultural Anthropology*, **25**: 545–576.

Kirschner, M., J. Gerhart and J. Mitchison. (2000). Molecular "vitalism". *Cell*, **100**: 79–88.

Kohn, E. (2007). How dogs dream: Amazonian natures and the politics of transgression. *American Ethnologist*, **34**: 3–24.
Koumandou, G. (1985). *Paradosis ikogeniakou dikeou* [Traditions of family law] (4th edition). Athens: Sakkoula.
Kronfeldner, M. E. (2006). Is cultural evolution Lamarckian? *Biology and Philosophy*, **22**: 493–512.
Kudielka, B. M. and C. Kirschbaum. (2005). Sex differences in HPA axis responses to stress: a review. *Biological Psychology*, **69**: 113–132.
Kuzawa, C. W. and E. Sweet. (2009). Epigenetics and the embodiment of race: developmental origins of US racial disparities in cardiovascular health. *American Journal of Human Biology*, **21**: 2–15.
Kuzuno, T. (1994). Ainu gods and beliefs. In *Circumpolar Religion and Ecology: an Anthropology of the North*, eds. T. Irimoto and T. Yamada. Tokyo: University of Tokyo Press, pp. 3–6.
Laland, K. L. and G. Brown. (2011). *Sense and Nonsense: Evolutionary Perspectives on Human Behavior*, 2nd edition. Oxford: Oxford University Press.
Laland, K. L., J. R. Kendal and G. Brown. (2007). The niche construction perspective: implications for evolution and human behavior. *Journal of Evolutionary Psychology*, **5**: 51–66.
Laland, K. L, J. Odling-Smee and S. Myles. (2010). How culture shaped the human genome: bringing genetics and the human sciences together. *Nature Reviews Genetics*, **11**: 137–148.
Landecker, H. (2003). Beginning and ending with apoptosis: cell death and biomedicine. In *Remaking Life and Death: Toward an Anthropology of the Biosciences*, eds. S. Franklin and M. Lock. Santa Fe, NM: SAR Press, pp. 23–59.
Landecker, H. (2007). *Culturing Life: How Cells Became Technologies*. Cambridge, MA: Harvard University Press.
Landecker, H. (2011). Food as exposure: nutritional epigenetics and the new metabolism. *BioSocieties*, **6**: 167–194.
Landesman, C. and R. Meeks. (2003). *Philosophical Skepticism*. Malden, MA: Blackwell.
Lansing, J. S. (2003). Complex adaptive systems. *Annual Review of Anthropology*, **32**: 183–204.
Latour, B. (1993). *We Have Never Been Modern*. Boston, MA: Harvard University Press.
Latour, B. (2004). *Politics of Nature: How to Bring the Sciences into Democracy*. Cambridge, MA: Harvard University Press.
Lave, J. (1988). *Cognition in Practice: Mind, Mathematics, and Culture in Everyday Life*. Cambridge: Cambridge University Press.
Lave, J. and E. Wenger. (1991). *Situated Learning: Legitimate Peripheral Participation*. Cambridge: Cambridge University Press.
Leach, E. R. (1970 [1954]). *Political Systems of Highland Burma: A Study of Kachin Social Structure*. London: Athlone Press.
Lehrman, D. S. (1953). A critique of Konrad Lorenz's theory of instinctive behavior. *Quarterly Review of Biology*, **28**: 337–363.
Lende, D. (2008). Poverty poisons the brain. In *Developmental Psychology, Embodiment, Human Variation, Inequality, Neural Plasticity, Stress*. Available at: http://neuroanthropology.net/2008/02/18/poverty-poisons-the-brain/.
Lévi-Strauss C. (1962). *La pensée sauvage*. Paris: Plon.
Lévi-Strauss C. (1963). The sorcerer and his magic. In *Structural Anthropology*. New York, NY: Basic Books, pp. 167–185.
Lévi-Strauss C. (1975 [1952]). *Race et histoire*. Paris: Noël Gonthier.
Levin, M. G. and L. P. Potapov. (1964). *The Peoples of Siberia*. Chicago, IL: University of Chicago Press.

Lévinas, E. (1990 [1967]). *Totality and Infinity: An Essay on Exteriority*. Pittsburgh, PA: Duquesne University Press.

Levinson, S. C. (2006). On the human "interaction engine". In *Roots of Human Sociality: Culture, Cognition and Interaction*, eds. N. Enfield and S. Levinson. Oxford: Berg, pp. 39–69.

Levinson, S. C. (2009). Language and mind: Let's get the issues straight! In *Making Sense of Language: Readings in Culture and Communication*, ed. S. D. Blum. Oxford: Oxford University Press, pp. 25–46.

Lewontin, R. C. (1978). Adaptation. *Scientific American*, **239**: 157–169.

Lewontin, R. C. (1982). Organism and environment. In *Learning, Development and Culture*, ed. H. C. Plotkin. Chichester: John Wiley and Sons.

Lewontin, R. C. (1983). The organism as the subject and object of evolution. *Scientia*, **118**: 63–82.

Ley, R. E., M. Hamady, C. Lozupone *et al.* (2008). Evolution of mammals and their gut microbes. *Science*, **320**: 1647–1651.

Liberski-Bagnoud, D. (2002). *Les dieux du territoire. Penser autrement la généalogie*. Paris: CNRS Éditions.

Lin, D. H. K. and E. R. Maher. (2010). DNA methylation: a form of epigenetic control of gene expression. *Obstetrician and Gynaecologist*, **12**: 37–42.

Lindee, S. and R. Ventura Santos. (2012). The biological anthropology of living human populations: world histories, national styles, and international networks. *Current Anthropology*, **53** (S5): 1–14.

Linhart, R. (1981). *The Assembly Line*. Amherst, MA: University of Massachusetts Press.

Lloyd, G. E. R. (2009). *Disciplines in the Making: Cross-Cultural Perspectives on Elites, Learning, and Innovation*. Oxford: Oxford University Press.

Lock, M. (1997). Decentering the natural body: making difference matter. *Configurations*, **5**: 267–292.

Lock, M. and V. K. Nguyen. (2010). *An Anthropology of Biomedicine*. Chichester: Wiley–Blackwell.

Loftus, A. (2009). The *Theses on Feuerbach* as a political ecology of the possible. *Area*, **41**: 157–166.

López-Jaramillo, P., L. Pradilla and Y. Bracho. (2005). Papel del adipocito en la inflamación del síndrome metabólico. *Acta Médica Colombiana*, **30**: 137–140.

Lumey, L. H. (1992). Decreased birth weights in infants after maternal in uterus exposure to the Dutch Famine of 1944–1945. *Pediatric Perinatal Epidemiology*, **6**: 240–253.

Lupton, D. (2003). *Medicine as Culture: Illness, Disease and the Body in Western Societies*. Thousand Oaks, CA: Sage.

Lythgoe, J. (1979). *The Ecology of Vision*. Oxford: Clarendon Press.

Maccoby, E. E. (2000). Parenting and its effects on children: on reading and misreading behavior genetics. *Annual Review of Psychology*, **51**: 1–27.

Mace, R., C. J. Holden and S. Shennan (eds.) (2005). *The Evolution of Cultural Diversity: A Phylogenetic Approach*. London: UCL Press.

Magnusson, D. and R. B. Cairns. (1996). Developmental science: toward a unified framework. In *Developmental Science*, eds. R. B. Cairns, J. G. H. Elder and E. J. Costello. Cambridge: Cambridge University Press, pp. 7–30.

Mair, V. H. and M. Bender. (2011). *The Columbia Anthology of Chinese Folk and Popular Literature*. New York, NY: Columbia University Press.

Marchand, T. H. J. (2001). *Minaret Building and Apprenticeship in Yemen*. Richmond: Curzon.

Marchand, T. H. J. (2003). A possible explanation for the lack of explanation; or 'Why the master builder can't explain what he knows': introducing the

informational atomism against a 'definitional' definition of concepts. In *Negotiating Local Knowledge: Power and Identity in Development*, eds. A. Bicker, P. Sillitoe and J. Pottier. London: Pluto Press, pp. 30–50.
Margulis, L. and D. Sagan. (2002). *Acquiring Genomes: A Theory of the Origins of Species*. New York, NY: Basic Books.
Marks, J. (2011). *The Alternative Introduction to Biological Anthropology*. Oxford: Oxford University Press.
Martin, E. (1990). Toward an anthropology of immunology: the body as nation state. *Medical Anthropology Quarterly*, **4**: 410–426.
Marx, K. (1973 [1857–8]). *Grundrisse*, transl. Martin Nicolaus. Harmondsworth: Penguin Books.
Marx, K. (1975). Economic and philosophical manuscripts. In *Karl Marx: Early Writings*, ed. L. Colletti. London: Penguin and New Left Review.
Marx, K. (1998 [1845]). Theses on Feuerbach. In *The German Ideology*, K. Marx with F. Engels. Amherst, NY: Prometheus Books.
Mateos, P. (2007). A review of name-based ethnicity classification methods and their potential in population studies. *Population Space and Place*, **13**: 243–263.
Maturana, H. and F. J. Varela. (1992). *The Tree of Knowledge: The Biological Roots of Human Understanding*. London: Shambhala.
Mauss, M. (1973 [1935]). Techniques of the body. *Economy and Society*, **2**: 70–88.
Mayr, E. (1999 [1942]). *Systematics and The Origin of Species from the Viewpoint of a Zoologist*. Cambridge, MA: Harvard University Press.
McGuire, W., A. V. S. Hill, C. E. M. Allsopp, B. M. Greenwood and D. Kwiatkowski. (1994). Variation in the TNF-alfa promoter region associated with susceptibility to cerebral malaria. *Nature*, **371**: 508–511
McLean, S. (2009). Stories and cosmogonies: imagining creativity beyond "nature" and "culture". *Cultural Anthropology*, **24**: 213–245.
Mcleod, J., C. J. Sinal and T. S. Perrot-Sinal. (2007). Evidence for non-genomic transmission of ecological information via maternal behavior in female rats. *Genes, Brain and Behavior*, **6**: 19–29.
Mensah, G. A., A. H. Mokdad, E. S. Ford, K. J. Greenlund and J. B. Croft. (2005). State of disparities in cardiovascular health in the United States. *Circulation*, **111**: 1233–1241.
Merleau-Ponty, M. (1968). *The Visible and the Invisible*, trans. A. Lingis. Evanston, IL: Northwestern University Press.
Mesoudi, A., D. Veldhuis and R. A. Foley. (2010). Why aren't the social sciences Darwinian? *Journal of Evolutionary Psychology*, **8**: 93–104.
Metho, O. O. (2006). *Anuak Justice Council* [Online]. Available at: http://www.anuakjustice.org/060624AddressToEthiopianCommunityInCalgary.htm [Accessed 27 September 2011].
Mollona, M. (2009). *Made in Sheffield: An Ethnography of Industrial Work and Politics*. Oxford: Berghahn Books.
Mollona, M., G. De Neve and J. Parry (eds.) (2009). *Industrial Work and Life: An Anthropological Reader*. Oxford: Berg.
Monod, J. (1970). *Le Hasard et la Nécessité*. Paris: Le Seuil.
Moore, M. K. (1992). Early imitation within a functional framework: the importance of person identity, movement and development. *Infant Behavior and Development*, **15**: 479–505.
Moss, L. (2001). Deconstructing the gene. In *Cycles of Contingency Developmental Systems and Evolution*, eds. S. Oyama, P. E. Griffiths and R. D. Gray. Cambridge, MA: MIT Press, pp. 85–98.
Moss, L. (2003). *What Genes Can't Do*. Cambridge, MA: MIT Press.

Müller, G. B. and S. A. Newman. (2005). The innovation triad: an EvoDevo agenda. *Journal of Experimental Zoology Part B: Molecular and Developmental Evolution*, **304**: 487–503.

Müller-Wille, S. and H. -J. Rheinberger. (2007). Heredity – the formation of an epistemic space. In *Heredity Produced: At the Crossroads of Biology, Politics, and Culture, 1500–1870*, eds. S. Müller-Wille and H.-J. Rheinberger. Cambridge, MA: MIT Press, pp. 3–34.

Nesbitt, S. and R. G. Victor. (2004). Pathogenesis of hypertension in African Americans. *Congestive Heart Failure*, **10**: 24–29.

Nettleton, S. (1995). *The Sociology of Health and Illness*. Cambridge: Polity Press.

Newton, T. (2010). Life times. *Sociological Review*, **58** (S1): 197–216.

Niemitz, E. L. and A. P. Feinberg. (2004). Epigenetics and assisted reproductive technology: a call for investigation. *American Journal of Human Genetics*, **74**: 599–609.

Niewöhner, J. (2011). Epigenetics: embedded bodies and the molecularisation of biography and milieu. *BioSocieties*, **6**: 279–298.

Noble, D. (2010). Biophysics and systems biology. *Philosophical Transactions of the Royal Society A: Mathematical, Physical and Engineering Sciences*, **368** (1914): 1125–1139.

Nowak, M. A. and R. Highfield. (2011). *Supercooperators: Altruism, Evolution, and Why We Need Each Other to Succeed*. New York, NY: Free Press.

Ochs, E. (1986). Introduction. In *Language Socialization Across Cultures*, eds. B. B. Schieffelin and E. Ochs. Cambridge: Cambridge University Press, pp. 1–13.

Ochs, E., R. C. Smith and C. E. Taylor. (1996). Detective stories at dinnertime: problem solving through co-narration. In *Disorderly Discourse: Narrative, Conflict, and Inequality*, ed. C. Briggs. New York, NY: Oxford University Press, pp. 95–113.

Odling-Smee, F. G. (1988). Niche-constructing phenotypes. In *The Role of Behavior in Evolution*, ed. H. C. Plotkin. Cambridge, MA: MIT Press, pp. 73–132.

Odling-Smee, F. J., K. N. Laland and M. W Feldman. (2003). *Niche Construction: The Neglected Process in Evolution*. Princeton, NJ: Princeton University Press.

Okamura, J. Y. (2008). *Ethnicity and Inequality in Hawai'i*. Philadelphia, PA: Temple University Press.

Olson, V. A. (2010). The ecobiopolitics of space biomedicine. *Medical Anthropology*, **29**: 170–193.

Oyama, S. (1985). *The Ontogeny of Information: Developmental Systems and their Evolution*. Cambridge: Cambridge University Press.

Oyama, S. (1992). Is phylogeny recapitulating ontogeny? In *Understanding Origins*, eds. F. J. Varela and J. P. Dupuy. Dordrecht: Kluwer Academic, pp. 227–232.

Oyama, S., P. E. Griffiths and R. D. Gray (eds.) (2001). *Cycles of Contingency: Developmental Systems and Evolution*. Cambridge, MA: MIT Press.

Pagel, M. (2012). *Wired for Culture: Origins of the Human Social Mind*. New York, NY: W. W. Norton & Company.

Pagli, C. and F. Sigmundsson. (2008). Will present day glacier retreat increase volcanic activity? Stress induced by recent glacier retreat and its effect on magmatism at the Vatnajökull ice cap, Iceland. *Geophysical Research Letters*, **35**: 1–5.

Palsson, G. (1994). Enskilment at sea. *Man* (N.S.), **29**: 901–927.

Palsson, G. (1996). Human–environmental relations. Orientalism, paternalism and communalism. In *Nature and Society: Anthropological Perspectives*, eds. P. Descola and G. Palsson. London: Routledge, pp. 63–81.

Palsson, G. (2006). Nature and society in the age of postmodernity. In *Reimagining Political Ecology*, eds. A. Biersack and J. B. Greenberg. Durham, NC: Duke University Press, pp. 70–93.

Palsson, G. (2007a). *Anthropology and the New Genetics*. Cambridge: Cambridge University Press.

Palsson, G. (2007b). How deep is the skin? The geneticization of race and medicine. *BioSocieties*, **2**: 257–272.

Palsson, G. (2008). Genomic anthropology: coming in from the cold? *Current Anthropology*, **49**: 545–568.

Palsson, G. (2009). Biosocial relations of production. *Comparative Studies in Society and History*, **51**: 288–313.

Palsson, G. (2010). Disciplining anthropology. *Norsk antropologisk tidskrift*, **21**: 198–210.

Palsson, G. (2012). Decode me!: anthropology and personal genomics. *Current Anthropology*, **53** (S5): 185–195.

Palsson, G. (2013). Life at the border: Nim Chimpsky *et al.* In *Nature and Anthropology*, ed. K. Hastrup. Abingdon: Routledge.

Palsson, G. and B. Prainsack. (2011). Genomic stuff: governing the (im)matter of life. *International Journal of the Commons*, **50**: 259–283.

Palsson, G., B. Szerszynski, S. Sörlin *et al.* (2013). Reconceptualizing the 'Anthropos' in the Anthropocene: integrating the social sciences and humanities in global environmental change research. *Environment Science and Policy*. http://dx.doi.org/10.1016/j.envsci.2012.11.004.

Pandya, V. (1993). *Above the Forest: A Study of Andamanese Ethnoanemology, Cosmology, and the Power of Ritual*. Delhi: Oxford University Press.

Pandya, V. (2009). *In the Forest: Visual and Material Worlds of Andamanese History*. Lanham, MD: University Press of America.

Panelli, R. (2010). More-than-human social geographies: posthuman and other possibilities. *Progress in Human Geography*, **34**: 79–87.

Papagaroufali, E. (2002). *Dora zois meta thanaton: politismikes ebiries* [Gifts of life after death: cultural experiences]. Athens: Ellinika Grammata.

Papanikolaou, E. and N. P. Anagnou. (2010). Major challenges for gene therapy of thalassemia and sickle cell disease. *Current Gene Therapy*, **10**: 404–412.

Papataxiarchis, E. and P. Loizos (eds.) (2001). *Contested Identities: Gender and Kinship in Modern Greece*. Princeton, NJ: Princeton University Press.

Pardini, M. and R. C. Guimaraes. (1992). A systemic concept of the gene. *Genetics and Molecular Biology*, **15**: 713–721.

Paredes, A., L. J. West and C. C. Snow. (1970). Biosocial adaptation and correlates of acculturation in the Tarahumara ecosystem. *International Journal of Social Psychiatry*, **16**: 163–174.

Parry, J. (1999). Lords of labour: working and shirking in Bhilai. *Contributions to Indian Sociology*, **33**: 107–140.

Paxson, H. (2004). *Making Modern Mothers: Ethics and Family Planning in Urban Greece*. Berkeley, CA: University of California Press.

Pearson, H. (2006). Genetics: what is a gene? *Nature*, **441**: 398–401.

Peltonen, L. and V. A. Mckusick. (2001). Genomics and medicine: dissecting human disease in the postgenomic era. *Science*, **291**: 1224–1229.

Pembrey, M. (2002). Time to take epigenetic inheritance seriously. *European Journal of Human Genetics*, **10**: 669–671.

Pembrey, M. E., L. O. Bygren, G. Kaati *et al.* (2006). Sex-specific, male-line transgenerational responses in humans. *European Journal of Human Genetics*, **14**: 159–166.

Perry, G. and R. Mace (2010). The lack of acceptance of evolutionary approaches to human behaviour. *Journal of Evolutionary Psychology*, **8**: 105–125.

Perthes, V. (ed.) (2004). *Arab Elites: Negotiating the Politics of Change*. Boulder, CO: Lynne Rienner Publishers.

Pigliucci, M., C. J. Murren and C. D. Schlichting. (2006). Phenotypic plasticity and evolution by genetic assimilation. *Journal of Experimental Biology*, **209**: 2362–2367.

Pina-Cabral, J. de. (2011). The truth about personal names. *Journal of the Royal Anthropological Institute*, **16**: 297–312.

Pinker, S. (2002). *The Blank Slate: The Modern Denial of Human Nature*. New York, NY: Viking Press.

Plessner, H. (1975 [1928]). *Die Stufen des Organischen und der Mensch: Einleitung in die philosophische Anthropologie*. Berlin: Walter de Gruyter.

Plomin, R. and H. Ho. (1991). Brain, behavior, and developmental genetics. In *Brain Maturation and Cognitive Development: Comparative and Cross-Cultural Perspectives*, eds. K. Gibson and A. C. Peterson. New York: Aldine de Gruyter, pp. 65–90.

Pocock, D. F. and J. MacClancy. (2001). *Understanding Social Anthropology*. London: Athlone Press.

Pottage, A. (2001). Persons and things: an ethnographic analogy. *Economy and Society*, **30**: 112–138.

Pottage, A. (2004). Introduction: The fabrication of persons and things. In *Law, Anthropology and the Constitution of the Social: Making Persons and Things*, eds. A. Pottage and M. Mundy. Cambridge: Cambridge University Press, pp. 1–39.

Praet, I. (2005). People into ghosts: Chachi death rituals as shape-shifting. *Tipiti: The Journal of the Society of Lowland South America*, **3**: 131–146.

Praet, I. (2006). *Courage and Fear: An Inquiry into Chachi Shape-Shifting*. Doctoral thesis, Oxford University.

Praet, I. (2009a). Shamanism and ritual in South America: an inquiry into Amerindian shape-shifting. *Journal of the Royal Anthropological Institute*, **15**: 737–754.

Praet, I. (2009b). Catastrophes and weddings: Chachi ritual as metamorphosis. *Journal de la Société des Américanistes*, **95**, 71–89.

Propper, C. and J. Rigg. (2007). Socio-economic status and child behaviour: evidence from a contemporary UK cohort (CASE Paper 125). London: London School of Economics/Centre for Analysis of Social Exclusion.

Proschan, F. (1997). "We are all Kmhmu, just the same": ethnonyms, ethnic identities, and ethnic groups. *American Ethnologist*, **24**, 91–113.

Rabinow, P. (1996). Artificiality and enlightenment: from sociobiology to biosociality. In *Essays on the Anthropology of Reason*. Princeton, NJ: Princeton University Press, pp. 91–111.

Rabinow, P. (2008). Afterword. In *Biosocialities, Genetics and the Social Sciences: Making Biologies and Identities*, eds. S. Gibbon and C. Novas. Abingdon: Routledge, pp. 188–192.

Radcliffe-Brown, A. R. (1922). *The Andaman Islanders*. New York, NY: Free Press.

Ramirez Goicoechea, E. (2009). *Evolución, Cultura y Complejidad. La Humanidad que se hace a sí misma* (2nd edition). Madrid: Editorial Universitaria Ramón Areces.

Rando, O. J. and K. J. Verstrepen. (2007). Timescales of genetic and epigenetic inheritance. *Cell*, **128**: 655–668.

Rappaport, R. (1968). *Pigs for the Ancestors: Ritual in the Ecology of a New Guinea People*. New Haven, CT: Yale University Press.

Reefhuis, J., M. A. Honein, L. A. Schieve *et al.* (2009). Assisted reproductive technology and major structural birth defects in the United States. *Human Reproduction*, **24**: 360–366.

Rheinberger, H. -J. (1995). Beyond nature and culture: a note on medicine in the age of molecular biology. *Science in Context*, **8**: 249–263.

Rheinberger, H. -J. (2010). *An Epistemology of the Concrete: Twentieth-Century Histories of Life*. Durham, NC: Duke University Press.

Richardson, M. (1982). Being-in-the-market versus being-in-the-Plaza: material culture and the construction of social reality in Spanish America. *American Ethnologist*, **9**: 421–436.

Richerson, P. J. and R. Boyd. (2008). *Not by Genes Alone: How Culture Transformed Human Evolution*. Chicago, IL: University of Chicago Press.

Rival, L. (2002). *Trekking Through History: The Huaorani of Amazonian Ecuador*. New York, NY: Columbia University Press.

Rizzolati, G. and C. Sinigaglia. (2006). *So quel que fai. Il cervello che agisce e i neuroni specchio*. Milano: Raffaelo Cortina Editore.

Robbins, J. and A. Rumsey. (2008). Cultural and linguistic anthropology and the opacity of other minds. *Anthropological Quarterly*, **81**: 407–420.

Robertson, A. F. (1996). The development of meaning: ontogeny and culture. *Man*, **4**: 591–610.

Rogoff, B. and J. Lave. (1984). *Everyday Cognition: Its Development in Social Context*. Cambridge, MA: Harvard University Press.

Rose, N. (2006). *The Politics of Life Itself: Biomedicine, Power, and Subjectivity in the Twenty-First Century*. Princeton, NJ: Princeton University Press.

Rose, N. (2007). Beyond medicalization. *Lancet* **369**: 700–771.

Rose, S., R. C. Lewontin and L. J. Kamin. (1976). *Not In Our Genes: Biology, Ideology and Human Nature*. Harmondsworth: Penguin Books.

Roseboom, T. J., J. H. Van Der Meulen, A. C. Ravelli *et al.* (2001). Effects of prenatal exposure to the Dutch famine on adult disease in later life: an overview. *Molecular Cell Endocrinology*, **185**: 93–98.

Roth, C. F. (2008). *Becoming Tsimshian: The Social Life of Names*. Seattle, WA: University of Washington Press.

Rothstein, M. A., Y. Cai and G. E. Marchant. (2009). The ghost in our genes: legal and ethical implications of epigenetics. *Health Matrix*, **19**: 1–62.

Rubenstein, S. (2002). *Alejandro Tsakimp: A Shuar Healer in the Margins of History*. Lincoln, NE: University of Nebraska Press.

Rubin, D. (1988). Go for the skill. In *Remembering Reconsidered: Ecological and Traditional Approaches to the Study of Memory*, eds. U. Neisser and E. Winograd. Cambridge: Cambridge University Press, pp. 374–382.

Salverte, E. (1864). *History of the Names of Men, Nations, and Places* (vol. II). London: Russell Smith.

Sapir, J. D. (1977). The anatomy of metaphor. In *The Social Use of Metaphor: Essays on the Anthropology of Rhetoric*, eds. J. D. Sapir and J. C. Crocker. Philadelphia, PA: University of Pennsylvania Press, pp. 3–32.

Sapolsky, R. M. (2004). Organismal stress and telomeric aging: an unexpected connection. *Proceedings of the National Academy of Sciences USA*, **101**: 17323–17324.

Sapolsky, R. M., L. M. Romero and A. U. Munck. (2000). How do glucocorticoids influence stress responses? Integrating permissive, suppressive, stimulatory, and preparative actions. *Endocrine Reviews*, **21**: 55–89.

Saussure, F. de. (1959 [1916]). *Course in General Linguistics*. New York, NY: McGraw-Hill.

Schabas, M. (2005). *The Natural Origins of Economics*. Chicago, IL: University of Chicago Press.

Scheff, T. J. (1988). Shame and conformity: the deference-emotion system. *American Sociological Review*, **53**: 395–406.

Schelling, F. W. Jv. (2004 [1799]). *First Outline of a System of the Philosophy of Nature*. Albany, NY: State University of New York Press.

Schieffelin, B. B. and E. Ochs (eds.) (1986). *Language Socialization Across Cultures*. Cambridge: Cambridge University Press.

Schneider, D. (1969). Kinship, nationality, and religion in American culture. In *Symbolic Action*, ed. V. Turner. Tulane, NC: American Ethnological Society, pp. 116–125.

Schultz, E. (2009). Resolving the anti-antievolutionism dilemma: a brief for relational evolutionary thinking in anthropology. *American Anthropologist*, **111**: 224–237.

Schwartz, J. and A. Pollack. (2010). Judge invalidates human gene patent. *The New York Times*, March 29.

Scott, J. C., J. Tehranian and J. Mathias. (2002). The production of legal identities proper to states: the case of the permanent family surname. *Comparative Studies in Society and History*, **44**: 4–44.

Scribner, S. and M. Cole. (1981). *The Psychology of Literacy*. Cambridge, MA: Harvard University Press.

Sear, R. and R. Mace. (2008). Who keeps children alive? A review of the effects of kin on child survival. *Evolution and Human Behaviour*, **29**: 1–18.

Segal, D. A. and S. J. Yanagisako. (2005). Introduction. In *Unwrapping the Sacred Bundle: Reflections on the Disciplining of Anthropology*, eds. D. A. Segal and S. J. Yanagisako. Durham, NC: Duke University Press, pp. 1–23.

Segerstrale, U. (2000). *Defenders of the Truth: The Battle for Science in the Sociobiology Debate and Beyond*. Oxford: Oxford University Press.

Sengupta, S. (2009). *The Tea Labourers of North East India: An Anthropo-Historical Perspective*. New Delhi: Mittal Publications.

Sgaramella, V. (2010). Variability of our somatic (epi)genomes. *Science*, **329**: 32–33.

Shao, H., L. C. Burrage, D. S. Sinasac *et al.* (2008). Genetic architecture of complex traits: large phenotypic effects and pervasive epistasis. *Proceedings of the National Academy of Sciences USA*, **105**: 19910–19914.

Shapiro, J. A. (1999). Genome system architecture and natural genetic engineering in evolution. In *Molecular Strategies in Biological Evolution*, ed. L. H. Caporale. New York, NY: New York Academy of Sciences.

Shapiro, J. A. (2005). A 21st century view of evolution: genome system architecture, repetitive DNA, and natural genetic engineering. *Gene*, **345**: 91–100.

Shapiro, J. A. and R. A. V. Sternberg. (2005). Why repetitive DNA is essential to genome function. *Biological Reviews*, **80**: 227–250.

Sharp, L. (2002). Bodies, boundaries, and territorial disputes: investigating the murky realm of scientific authority. *Medical Anthropology*, **21**: 369–379.

Sheridan, M. J. (2008). The dynamics of African sacred groves: ecological, social and symbolic processes. In *African Sacred Groves: Ecological Dynamics and Social Change*, ed. M. J. Sheridan and C. Nyamweru. Oxford: James Currey, pp. 9–41.

Shook, R. P., L. Duck-shul, S. Xuemei *et al.* (2012). Cardiorespiratory fitness reduces the risk of incident hypertension associated with a parental history of hypertension. *Hypertension*. Published online before print 14 May, 2012, doi: 10.1161/HYPERTENSIONAHA.112.191676.

Shore, B. (1996). *Culture in Mind: Cognition, Culture and the Problem of Meaning*. Oxford: Oxford University Press.

Shore, C. (2010). Beyond the multiversity: neoliberalism and the rise of the schizophrenic university. *Social Anthropology*, **18**: 15–29.

Simpson, G. G. (1953). The Baldwin effect. *Evolution*, **7**: 110–117.

Simpson, J. H., D. Nash, M. Laughren, P. Austin and B. Alpher (2001). *Forty Years On: Ken Hale and Australian Languages*. Canberra: Research School of Pacific and Asian Studies, Australian National University.

Smith, A. B. (2002). *Where Have All the Hottentots Gone? The Archaeology & History of the Khoekhoen* [Online]. Available at: http://www.scienceinafrica.co.za/2002/august/khoi.htm [Accessed 15 July 2011].

Smith, H. (2008). *Portrait of a People: the Tiwi of Northern Australia*. Narooma, New South Wales: Hobbs Point Publishing.

Smith, P. (2002). *Modern English-Mienh and Mienh-English Dictionary*. Victoria, BC: Trafford Publishing.

Smocovitis, V. B. (2012). Anthropology, the evolutionary synthesis, and the prehistory of biological anthropology, 1927–1962. *Current Anthropology*, **53** (S5): 108–125.

Sofaer, J. R. (2006). *The Body as Material Culture: A Theoretical Osteoarchaeology*. Cambridge: Cambridge University Press.

Sperber, D. (1996). *Explaining Culture: A Naturalistic Approach*. Oxford: Blackwell.

Standen, V. and R. A. Foley (eds.) (1989). *Comparative Socioecology: The Behavioural Ecology of Humans and Other Mammals*. Oxford: Blackwell.

Steiner, G. (1975). *After Babel*. Oxford: Oxford University Press.

Stotz, A. (2006). With genes like that, who needs an environment? Postgenomics' argument for the ontogeny of information. *Philosophy of Science*, **73**: 905–917.

Stotz, A. and P. E. Griffiths. (2008). Biohumanities: rethinking the relationship between biosciences, philosophy and history of science, and society. *Quarterly Review of Biology*, **83**: 37–45.

Stotz, K. (2008). The ingredients for a postgenomic synthesis of nature and nurture. *Philosophical Psychology*, **21**: 359–381.

Strathern, M. (1991). One man and many men. In *Big Men and Great Men: Personifications of Power in Melanesia*, eds. M. Godelier and M. Strathern. Cambridge: Cambridge University Press, pp. 197–214.

Strathern, M. (1992a). *After Nature: English Kinship in the Late-Twentieth Century*. Cambridge: Cambridge University Press.

Strathern, M. (1992b). *Reproducing the Future: Anthropology, Kinship and the New Reproductive Technologies*. Manchester: University of Manchester Press.

Strathern, M. (1996). The concept of society is theoretically obsolete. In *Key Debates in Anthropology*, ed. T. Ingold. London: Routledge, pp. 49–55.

Strathern, M. (1998). Divisions of interest and the languages of ownership. In *Property Relations: Renewing the Anthropological Tradition*, ed. C. Hann. Cambridge: Cambridge University Press, pp. 214–232.

Strathern, M. (1999). *Property, Substance and Effect*. London: Athlone Press.

Strathern, M. (2005). Divided origins and the arithmetic of ownership. In *Kinship, Law and the Unexpected: Relatives are Always a Surprise*. Cambridge: Cambridge University Press, pp. 135–162.

Sturtevant, W. C. (ed.) (1978). *Handbook of North American Indians* (vol. 15, Northeast). Washington, DC: Smithsonian Institution Press.

Sturtevant, W. C. (ed.) (1998). *Handbook of North American Indians* (vol. 12, Plateau). Washington, DC: Smithsonian Institution Press.

Susser, M. and Z. Stein. (1994). Timing in prenatal nutrition: a reprise of the Dutch famine study. *Nutrition Reviews*, **52**: 84–94.

Sussman, R. W. and A. R. Chapman (eds.) (2004). *The Origins and Nature of Sociality*. Hawthorne, NY: de Gruyter.

Sussman, R. W. and R. C. Cloninger. (2011). *Origins of Altruism and Cooperation. Developments in Primatology: Progress and Prospects*, Vol. 36, Part 2. New York, NY: Springer.

Szerszynski, B. (2003). Technology, performance and life itself: Hannah Arendt and the fate of nature. *Sociological Review*, **51**: 203–218.

Taher, A. T., K. M. Musallam and M. D. Cappellini. (2009). Thalassaemia intermedia: an update. *Mediterannean Journal of Hematology and Infectious Diseases* **1**. Available at: http://www.mjhid.org/article/view/4709/html_6.

Taher, A. T., K. M. Musallam, M. D. Cappellini and D. J. Weatherall. (2011). Optimal management of β thalassaemia intermedia. *British Journal of Hematology*, **152**: 512–523.

Talbot, N. L., B. Chapman, C. Yeates *et al.* (2009). Childhood sexual abuse is associated with physical illness burden and functioning in psychiatric patients 50 years of age and older. *Psychosomatic Medicine*, **71**: 417–422.

Templeton, A. (1998). Human races: a genetic and evolutionary perspective. *American Anthropologist*, **100**: 632–650.

Thavagnanam, S., J. Fleming, A. Bromley, M. D. Shields and C. R. Cardwell. (2008). A meta-analysis of the association between Caesarean section and childhood asthma. *Clinical and Experimental Allergy*, **38**: 629–633.

Theise, N. D. and R. Harris. (2006). Postmodern biology: (adult) (stem) cells are plastic, stochastic, complex, and uncertain. In *Stem Cells, Handbook of Experimental Pharmacology*, eds. A. M. Wobus and K. Boheler. New York, NY: Springer, pp. 389–408.

Thompson, W. E. (1983). Hanging tongues: a sociological encounter with the assembly line. *Qualitative Sociology*, **6**: 215–237.

Thoreau, H. D. (1906). *The Writings of Henry David Thoreau*. Boston, MA: Houghton Mifflin.

Thrift, N. (2004). Bare life. In *Cultural Bodies: Ethnography and Theory*, eds. H. Thomas and J. Ahmed. Oxford: Blackwell, pp. 145–169.

Tomasello. M. (2008). *Origins of Human Communication*. Cambridge, MA: MIT Press.

Tomasello, M. and M. Carpenter. (2007). Shared intentionality. *Developmental Science*, **10**: 121–125.

Tomasello, M. and H. Rakoczy. (2003). What makes human cognition unique? From individual to shared to collective intentionality. *Mind and Language*, **18**: 121–147.

Tooby, J. and L. Cosmides. (1992). The psychological foundations of culture. In *The Adapted Mind: Evolutionary Psychology and the Generation of Culture*, eds. J. H. Barkow, L. Cosmides and J. Tooby. New York, NY: Oxford University Press, pp. 19–136.

Traka, D. (2007). *I iatriki sti sighroni Ellada* [Medicine in contemporary Greece]. Arhaeology 105, Athens.

Trivers, R. (1971). The evolution of reciprocal altruism. *Quarterly Review of Biology*, **46**: 35–57.

Turnbaugh, P. J., R. E. Ley, M. Hamady *et al.* (2007). The Human Microbiome Project. *Nature*, **449**: 804–810.

Turner V. W. (1974). *Dramas, Fields and Metaphors: Symbolic Action in Human Society*. Ithaca, NY: Cornell University Press.

Turner V. W. (1995 [1969]). *The Ritual Process: Structure and Anti-Structure*. New York, NY: Aldine de Gruyter.

Tusting, K. (2005). Language and power in communities of practice. In *Beyond Communities of Practice: Language, Power, and Social Context*, eds. D. Barton and K. Tusting. Cambridge: Cambridge University Press, pp. 36–54.

Ukpolo, C. (2010). Ezigbo Mmadu: an anthropological investigation into the concept of a good person in Igbo worldview. *Lumina*, **21**: 1–13.

Van Den Berg, G. J., P. Pinger and J. Schoch. (2011). Instrumental variable estimation of the causal effect of hunger early in life on health later in life. *IZA Discussion Paper*, No. 6110.

Vansina, J. (1990). *Paths in the Rainforest: Toward a History of Political Tradition in Equatorial Africa*. Madison, WI: University of Wisconsin Press.

Varese, S. (1968). *Salt of the Mountain*. Norman, OK: University of Oklahoma Press.

Venter, J. C., M. D. Adams, E. W. Myers *et al.* (2001). The sequence of the human genome. *Science*, **291** (5507): 1304–1351.

Ventura, M. (2009). *Identité, cosmologie et chamanisme des Tsachila de l'Equateur: A la croisée des chemins*. Paris: L'Harmattan.

Vilaça, A. (2006). *Quem somos nós: Os Wari' encontram os brancos*. Rio de Janeiro: Editora UFRJ.

Vilaça, A. (2007). Cultural change as body metamorphosis. In *Time and Memory in Indigenous Amazonia: Anthropological Perspectives*, eds. C. Fausto and M. J. Heckenberger. Gainesville: University Press of Florida, pp. 169–193.

Viveiros de Castro, E. (1998). Cosmological deixis and Amerindian perspectivism. *Journal of the Royal Anthropological Institute*, **4**: 469–488.

Viveiros de Castro, E. (2004). Exchanging perspectives: the transformation of objects into subjects in Amerindian ontologies. *Common Knowledge*, **10**: 463–484.

Viveiros de Castro, E. (2012). Cosmological perspectivism in Amazonia and elsewhere. *HAU Masterclass Lectures*, **1**: 45–168.

Volpi, F. (1994). Being and time: A "translation" of the Nichomachean Ethics? In *Reading Heidegger From the Start: Essays in His Earliest Thought*, eds. T. J. Kisiel and J. Van Buren. Albany, NY: State University of New York Press, pp. 195–211.

Volpi, F. (1996). Dasein as *praxis*: the Heideggerian assimilation and the radicalization of the practical philosophy of Aristotle. In *Critical Heidegger*, ed. C. Macann. London: Routledge, pp. 27–66.

Von Uexküll, J. (1982). The theory of meaning. *Semiotica*, **42**: 25–82.

Waddington, C. H. (1942). Canalization of development and the inheritance of acquired characters. *Nature*, **150**: 563–565.

Waddington, C. H. (1957). *The Strategy of the Genes: A Discussion of Some Aspects of Theoretical Biology*. London: George Allen and Unwin.

Wagner, W., N. Kronberger, N. Allum *et al.* (2002). Pandora's genes – images of genes and nature. In *Biotechnology: The Making of a Global Controversy*, eds. M. W. Bauer and G. Gaskell. Cambridge: Cambridge University Press, pp. 244–276.

Waldman, C. (2006). *Encyclopedia of Native American tribes*. New York, NY: Infobase Publishing.

Walls, P. and R. Williams. (2004). Accounting for Irish Catholic ill health in Scotland: a qualitative exploration of some links between "religion", class, and health. *Sociology of Health and Illness*, **26**: 527–556.

Watson, J. D. and F. H. Crick. (1953). Genetical implications of the structure of deoxyribonucleic acid. *Nature*, **171**: 964–967.

Weatherall, D. J. (1998). Thalassaemia in the next millennium. *Annals of the New York Academy of Sciences*, **850**: 1–9.

Weatherall, D. J. (2001). Phenotype-genotype relationship in monogenic disease: lessons from the Thalassaemias. *Nature Reviews/Genetics*, **2**: 245–255.

Weatherall, D. J. and J. B. Clegg. (2001). *The Thalassaemia Syndromes*, 4th edition. Oxford: Blackwell Science.

Weatherall, D. J, J. B. Clegg and D. Kwiatkowsi. (1997). The role of genomics in studying genetic susceptibility to infectious disease. *Genome Research*, **7**: 967–973.

Weiss, B. (1996). *The Making and Unmaking of the Haya Lived World: Consumption, Commoditization, and Everyday Practice*. Durham, NC: Duke University Press.

Wenger, E. (1998). *Communities of Practice: Learning, Meaning and Identity*. Cambridge: Cambridge University Press.

Werbner, R. (ed.) (2002). *Postcolonial Subjectivities in Africa*. London: Zed Books.

West-Eberhard, M. J. (2003). *Developmental Plasticity and Evolution*. Oxford: Oxford University Press.

West, M. J. and A. P. King. (2008). Deconstructing innate illusions: reflections on nature–nurture–niche from an unlikely source. *Philosophical Psychology*, **21**: 383–395.

White, T. D., B. Asfaw, Y. Beyene, Y. Haile-Selassie, C. O. Lovejoy, G. Suwa and G. WoldeGabriel. (2009). *Ardipithecus ramidus* and the paleobiology of early hominids. *Science*, **326**: 75–86.

Whitehead, A. N. (1926). *Science and the Modern World*. New York, NY: Free Press.

Wiklund, H., A. G. Glover, P. J. Johannessen and T. G. Dahlgren. (2009). Cryptic speciation at organic-rich marine habitats: a new bacteriovore annelid from whale-fall and fish farms in the North-East Atlantic. *Zoological Journal of the Linnean Society*, **155**: 774–785.

Willerslev, R. (2007). *Soul Hunters: Hunting, Animism, and Personhood Among the Siberian Yukaghirs*. Berkeley, CA: University of California Press.

Wilson, E. O. (1975). *Sociobiology: The New Synthesis*. Cambridge, MA: Harvard University Press.

Wimsatt, W. and J. C. Schank. (1986). Generative entrenchment and evolution. In *PSA: Proceedings of the Biennial Meeting of the Philosophy of Science Association, Symposia and Invited Papers*, pp. 33–60.

Witherspoon, G. (1997). *Navaho Kinship and Marriage*. Chicago, IL: Chicago University Press.

Wittgenstein, L. (1967). *Zettel*. Oxford: Blackwell.

Wittgenstein, L. (2007 [1953]). *Philosophical Investigations*. Oxford: Blackwell.

Wood, B. and N. Lonergan. (2008). The fossil record: taxa, grades and clades. *Journal of Anatomy*, **212**: 354–376.

Worster, D. (1994), *Nature's Economy: A History of Ecological Ideas*. Cambridge: Cambridge University Press.

Yajnik, C. S. (2004). Early life origins of insulin resistance and type 2 diabetes in India and other Asian countries. *Journal of Nutrition*, **134**: 205–210.

Yanagisako, S. and C. Delaney. (1995). Naturalizing power. In *Naturalizing Power: Essays in Feminist Cultural Analysis*, eds. S. Yanagisako and C. Delaney. New York, NY: Routledge, pp. 1–22.

Yannaki, E., D. Emery, and G. Stamatoyannopoulos. (2010). Gene therapy for β-thalassaemia: The continuing challenge. *Expert Reviews in Molecular Medicine*, **12**: e31.

Yannaki, E., T. Papayannopoulou, E. Jonlin *et al.* (2012). Hematopoietic stem cell mobilization for gene therapy of adult patients with severe β-thalassaemia: Results of clinical trials using G-CSF pr plerixafor in splenectomized and nonsplenectomized subjects. *Molecular Therapy*, **20**: 230–238.

Yehuda, R. and L. M. Bierer. (2008). Transgenerational transmission of cortisol and PTSD risk. *Progress in Brain Research*, **167**: 121–135.

Yoshimura, M. (2007). *Weaving and Identity of the Atayal in Wulai, Taiwan*. Waterloo, ON: Waterloo University Press.

Zerilli, L. M. G. (1995). The Arendtian body. In *Feminist Interpretations of Hannah Arendt*, ed. B. Honig. University Park, PA: Pennsylvania State University Press, pp. 167–193.

Zizek, S. (2001). *On Belief*. London: Routledge.

Notes on the contributors

Hayder Al-Mohammad is currently the Sociological Review Fellow at the University of Keele. He has carried out more than two years of fieldwork in Basra and Baghdad between 2005 and 2012, focusing on the violence and deprivation of post-invasion Iraq.

Aglaia Chatjouli holds a PhD in Social Anthropology (University of the Aegean) and has a background in Biology. She has carried out fieldwork in Greece studying thalassaemia and the cultural construction of genetic illness. She has worked on the public understanding of biotechnology in Europe and on migrants' health. Her research concerns the interface between biology and anthropology, the relation between (bio)difference and normality and imaginings of nature.

Bob Forrester is archaeological consultant to the Swaziland National Museum. He created a digital database of San art with the Rock Art Research Institute of the University of the Witwatersrand. He is currently recording all archaeological sites in Swaziland and covering the country as a documentary photographer. He founded Swaziland Digital Archives to preserve Swaziland's photographic heritage. He recently redesigned and rebuilt all displays at the Swaziland National Museum.

Agustín Fuentes is Professor of Anthropology at the University of Notre Dame. His current research includes cooperation and community in human evolution, ethnoprimatology and multispecies anthropology, evolutionary theory and interdisciplinary approaches to human nature(s). Fuentes' recent books include *Evolution of Human Behavior* (Oxford) and *Race, Monogamy, and Other*

Lies They Told You: Busting Myths About Human Nature (University of California).

Barbara Götsch is a Researcher at the Institute for Social Anthropology (ISA) of the Austrian Academy of Sciences (AAS) in Vienna, employed in the doctoral programme of the arts and humanities section of the AAS. She has carried out ethnographic fieldwork in urban Morocco and is interested in cooperation and conflict, knowledge and communication, and more generally in the socio-cognitive aspects of life.

Tim Ingold is Professor of Social Anthropology at the University of Aberdeen. He has carried out ethnographic fieldwork in Lapland, and has written on environment, technology and social organization in the circumpolar North, on evolutionary theory, human–animal relations, language and tool use, environmental perception and skilled practice. He is currently exploring issues on the interface between anthropology, archaeology, art and architecture.

Vito Laterza recently completed his PhD in Social Anthropology at the University of Cambridge and is currently a Postdoctoral Fellow in the Human Economy Programme, University of Pretoria. His doctoral work explores foundational issues of logic, belief and experience through an ethnography of timber workers in Swaziland. At present, he is developing a nondualist framework for the cross-cultural study of these topics.

Gaetano Mangiameli (PhD University of Bologna, 2006) is Lecturer in Cultural Anthropology at the University of Bologna. He has carried out fieldwork in the Upper East Region of Ghana, in Sicily and in Northern Italy and has written on environmental issues, complexity and emergence, the sacralization of nature, funerary practices, human–animal relations and tourism. His further fields of interest are urban agriculture, mass communication, new media and social movements.

Patience Mususa is a PhD candidate in Social Anthropology at the University of Cape Town. She has carried out fieldwork on the Zambian Copperbelt exploring the texture of place and the experience of middle-class decline. She has practised in architecture, and has a broad interest in the anthropology of the built environment.

Gisli Palsson is Professor of Anthropology, University of Iceland, Reykjavik, Iceland. He has written on a variety of issues, including biomedicine, genomics, genetic history, human-animal relations, fishing, arctic exploration, environmental discourse and the history of slavery. He has carried out anthropological fieldwork in Iceland, the Republic of Cape Verde and the Canadian Arctic.

Istvan Praet is Senior Lecturer in Social Anthropology at the University of Roehampton (London). His principal research interest is the comparative study of cultural and scientific conceptions of life. He has conducted ethnographic fieldwork in Ecuador, where he studied Chachi animism, notions of metamorphosis and the perception of catastrophes. He has an emerging interest in the anthropology of space exploration and astrobiology.

Eugenia Ramirez-Goicoechea is Habilitated Professor of Social Anthropology at UNED, Madrid. She has carried out fieldwork in the Basque Country on ethnic identities and in Spain on international migrations, focusing on the political economy of social classifications. She has explored the potential of dynamic systems theories and complexity in the social sciences. Currently, she is working on a biosocial/biocultural approach to human evolution and ontogeny.

Noa Vaisman is an International Junior Research Fellow and a Marie Curie Fellow at Durham University. She conducted ethnographic fieldwork in Buenos Aires, Argentina and has written on such topics as human rights, new technologies and the body, sociality and temporality. She is currently finishing a book on the identification of the children of the disappeared in Argentina.

Index